Analysis of Engineering Drawings and Raster Map Images

Thomas C. Henderson

Analysis of Engineering Drawings and Raster Map Images

Thomas C. Henderson
University of Utah
Salt Lake City, UT, USA

ISBN 978-1-4939-5178-9 ISBN 978-1-4419-8167-7 (eBook)
DOI 10.1007/978-1-4419-8167-7
Springer New York Heidelberg Dordrecht London

Softcover reprint of the hardcover 1st edition 2014

Printed on acid-free paper

Springer is part of Springer Science+Business Media (www.springer.com)

Preface

Much computer vision research focuses on the physics of vision or image formation, as well as the analysis and understanding of natural scenes. The work here, on the other hand, deals strictly with human artifacts: engineering drawings and maps. Although electronic artifacts dominate our world today, there remain many legacy documents which require automated analysis. For example, many times in the past only paper drawings were acquired for vehicles and parts, and the only way to make them accessible is to digitize and catalog them for current users. Given that these documents can number in the thousands, this is too much to accomplish by hand.

As for maps, these too exist in profuse numbers, including historical documents, which motivates their automated analysis. In addition, it is often useful to register these with digital imagery, and the discovery of semantic features is essential for this task.

The search for algorithms to automate such analysis has been underway since the beginning of digital image processing, and progress has been steady in attaining the level of performance of today's systems. This book describes the state of the art in engineering drawing and raster map analysis and provides a starting point for future research in this area.

Salt Lake City, UT, USA
Thomas C. Henderson

Acknowledgments

I would like to thank all those who participated in developing the ideas and systems presented in this book which describes the last decade of our work. This includes the Alpha_1 research team with whom I have had the pleasure of working for many years, and more recently on the Viper reverse engineering project; many thanks to Elaine Cohen, Rich Riesenfeld, Jim de St. Germain, and David Johnson for interesting research and discussion on reverse engineering. I would also like to thank the colleagues at IAVO, namely Eric Lester and Brad Grinstead, with whom I worked on raster map image analysis. Throughout all this work there was a steady flow of student colleagues, including Lavanya Swaminathan, Chimiao Xu, Trevor Linton, Sergey Potupchik, Andrei Ostani, Anshul Joshi, Srivishnu Satyavolu, Xiuyi Fan, and Wenyi Wang; these students worked on various aspects of the projects mentioned above. The support through ARO grant number DAAD19-01-1-0013 and the AFOSR grant FA9550-08-C-0005 was also much appreciated. Finally, I would like to thank my wife, Joëlle, who has had to put up with my many hours spent at the computer rather than hiking and skiing with her!

Contents

Chapter 1
Introduction

The semantic analysis of human produced image artifacts is a difficult task. This includes raster images of engineering (CAD) drawings and geographic maps. Although these are very distinct problems, there are a number of common issues which must be addressed albeit from different viewpoints. We hope to bring to light alternative approaches to thinking about and solving these problems, and in that way, allow more general abstract methods to emerge.

1.1 Engineering Drawing Analysis

The automatic semantic interpretation of engineering drawings is primarily a document image analysis problem [119]; for a summary of this research area, see [118]. To get an idea of the scale of the problem, it has been reported that about 250 million drawings are generated annually (see [62]). An important aspect of this is performance analysis; for more see [80] where document image segmentation algorithms are carefully compared. Related work includes that on interesting and challenging research problems in document analysis, pattern recognition, image processing and computer vision (see [60, 113, 120, 121, 138]).

Legacy engineering drawings such as that shown in Fig. 1.1 exist in profuse numbers without the associated electronic CAD data; e.g., the U.S. Army has tens of thousands of scanned engineering drawing images with little means of automatically accessing and exploiting them. Typically, the goal is to use the information in these drawings to make modifications of existing systems, or to make a new design derived from the old; one may also be interested in producing a 3D model from the 2D views given in the drawing. During the reverse engineering process, it may be possible to take advantage of an existing part, as well as any available drawings; oftentimes, an existing part will be broken or worn, and it is necessary to get correct dimension specifications from the drawing as well. Our major goal in the work we have performed is to segment the image as accurately as possible in order to permit the most correct interpretation of the annotations in the digitized drawing.

T.C. Henderson, *Analysis of Engineering Drawings and Raster Map Images*,
DOI 10.1007/978-1-4419-8167-7_1,

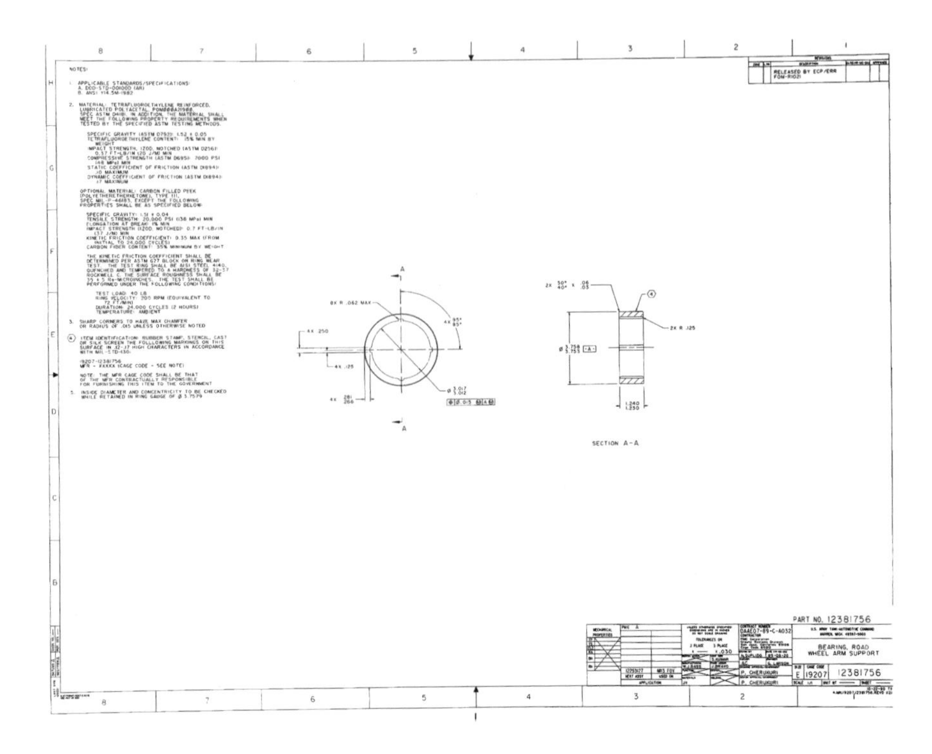

Fig. 1.1 A legacy engineering drawing

Although much progress has been made in specific areas of engineering drawing analysis, such as vectorization, text/graphics segmentation, etc., there are still no complete high performance analysis systems. Existing systems tend to be brittle in some aspect when applied to real-world scanned images. The focus of most previous work has been on straight-line vectorization [40, 83, 108], arc segmentation [109], dimensioning analysis [29, 111], and graphics analysis [7, 74].

Engineering drawing analysis usually starts with a black and white (binary) scanned image of the drawing under consideration (see Baird [13] for background structure analysis). Note that scanned raster images of maps tend to be color images, and require a somewhat different starting point. The image processing sequence is shown in Fig. 1.2. The recovery of connected components is a solved problem (e.g., Matlab provides the *bwlabel* function), and we assume that connected components are available in an image and distinctly numbered from 1 to *max_cc*. We have seen in Fig. 1.1 the type of engineering drawing image from the set we will analyze. The left side of Fig. 1.3 shows a sub-image of the legend area of the drawing, and on the right is shown the connected component corresponding to the line structure of the legend form.

The vectorization of the components is the next important step in the process. Many methods have been proposed for this, and generally include some form of

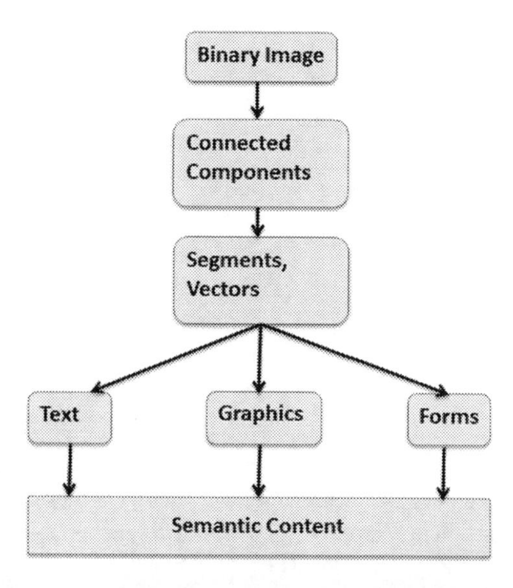

Fig. 1.2 The engineering drawing analysis process

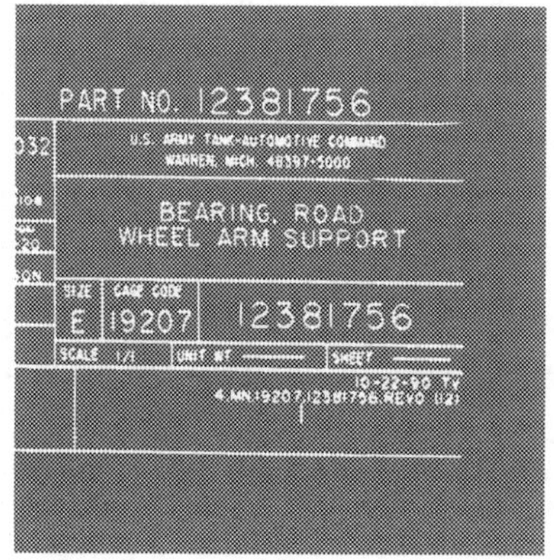

Blowup of Legend Area of the Engineering Drawing

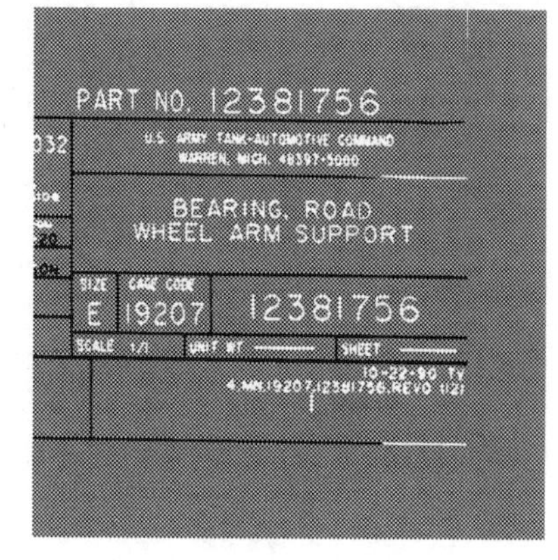

Connected Component 1 is in Black

Fig. 1.3 An example connected component

thinning or skeletonization. As pointed out by O'Gorman and Kasturi [92], this requires that:

- the number of connected components remains the same,
- end locations of segments exist in the skeleton,
- the resulting skeletons be roughly centered in the component, and
- that extraneous spurs be minimized.

For example, consider the character "T" shown in Fig. 1.4. The Matlab function *bwmorph* can be used to obtain the skeleton shown overlayed on the "T" in Fig. 1.4. As can be seen, the standard skeleton operator usually produces spurs that run to

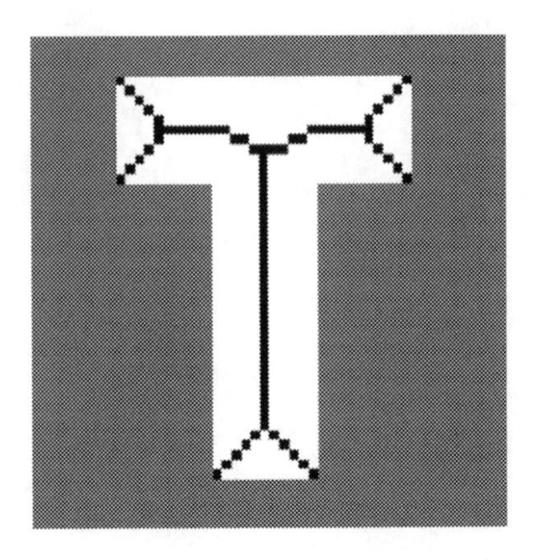

Fig. 1.4 The skeleton produced by Matlab's *bwmorph* function

rectangular corners. We will describe methods in later chapters which avoid this problem. Another interesting approach suggested by Dori [30], is the *Orthogonal Zig-Zag Method.* The main idea is to conceptually do a ray trace through the foreground line drawing as if shining light and letting it reflect when it hits the background. This is an excellent insight in that it leads to the application of a continuous process to the digital analysis problem. This idea is exploited in our method described in the next chapter. Shen et al. [106] introduced the Bending Potential Ratio (BPR) as a measure to help prune spurious shoots off the skeleton. Both local and global shape context are taken into consideration, and the resulting skeleton is generally better located along the true medial axis and produces more natural results. More recently, Mandal et al. [79] proposed a method which uses straightness properties to classify primitives into horizontal, vertical, inclined, and curved segments. An efficient way to do this is given, and experimental performance results yield about an 85% recognition accuracy. Another interesting approach is that of Xu [131] who used robot localization techniques to produce skeletons by "driving" through the line segments.

Once a robust skeleton is found, the next step is to vectorize the sequence of pixels into a set of polylines (also called *polygonalization*). Note that chain codes may be produced, and these form the lowest level of vectorization possible. Vectors, curved segments, etc. are produced from the skeletons, and this involves line and curve fitting. Once these are available, various types of shape analysis tools can be used to classify components into text characters, forms and graphics. For example, Song et al. [110] introduced an object-oriented progressive simplification vectorization scheme. This method attempts to use intrinsic characteristics to recognize classes of graphic objects. The workflow of the method extracts entities in the sequence: straight line segments (*Bar Vectorization*), arcs (*Arc Vectorization*), curves (*Curve Vectorization*), symbols (*Symbol Vectorization*), and text (*Text Vectorization*). The authors provide performance results of around 93% recognition rates on these categories.

An interesting technique for graphics vectorization is given by Hilaire and Tombre [53] which produces a skeletonization, and then segments it into primitives which are most likely while preventing over-fragmentation. The segmentation is based on the best fit of primitive arcs to segment points. It also has some special rules to remove false primitives and produces good positions on branch points (also called junction points). Overall, the method is quite robust and can be applied to a wide range of types of graphical drawings. For more on geometric reconstruction from engineering drawings, see [7,74]; however, we do not address this issue in this book.

Liu and Dori [75] proposed a generic graphic object recognition approach in which structural rules are checked during a flexible and adaptive component grouping process. First, vectors are produced by their sparse pixel tracking method [73]; this is a two-stage process, generating a coarse set of vectors, followed by a more refined set. The analysis is guided by two main principles: (1) shape preservation, and (2) efficiency. The graphic component recognition is achieved through a hypothesize and test approach. An example image result is given in the paper, and the authors state that their method was the top performer at a graphics recognition competition, but no details are given. Liu et al. [75] further extended their graphics recognition methods with an interactive approach. An instance of the particular graphic object is given to the system by the user, and a structural model for that object is learned in terms of geometric constraints. This structure is then sought among segmented objects from an image.

Many techniques exist for character and symbol recognition in general documents, as well as in engineering drawings specifically (e.g., see [33,35,55,78,115, 122]; also note that Lu [78] proposed that non-text objects be separated first, and that what remained be analyzed as possible text; this allows language generality—both Western and Chinese characters recognized, and is robust to characters touching graphics, font, orientation, and noise). One of the most influential papers on this topic is that of Fletcher and Kasturi [35]. In that work, a number of constraints are imposed:

- no more than a 5X scale difference can exist between different fonts,
- the spacing between lines must be greater than $\frac{1}{4}$ the character height,
- characters do not touch,
- the area of the largest characters is less than 5X the average component area, and
- the gap between characters is small.

An algorithm is then given which can locate and separate text with three or more characters; this involves:

- Find bounding box for each connected component.
- Apply an area/ratio filter. The paper uses two thresholds:
  - $T_1 = 5(A_{mp}, A_{avg})$, where A_{mp} is the number of components in the maximum density area, and A_{avg} is the number of components in the average component area.
  - T_2: the maximum elongation for text.

- Classify as text all components with an area less than T_1, a $\frac{height}{width} \in [\frac{1}{T_2}, T_2]$, and both height and width less than $\sqrt{T_1}$; others are classified as graphics.
- Group collinear components. The Hough transform is used for this, and text spacing helps produce word hypotheses.

Although this approach scales well in applying the connected component size filtering process, issues arise when dealing with drawings that have short character sequences or that do not satisfy the constraints. In addition, small components (e.g., commas, periods, dashes, etc.) may confuse the statistics.

A subsequent paper by Tombre et al. [122] expands on the Fletcher-Kasturi algorithm with a number of very useful extensions. First, they note that in engineering drawing and map analysis, an absolute threshold on character size can work well. They propose a separate shape filtering mechanism which separates the components into small (text), large (graphics), and a third category of components is introduced: the set of small elongated components. This latter category can be either graphics or text. A more flexible method for string detection is proposed which exploits a best fitting rectangle and they also allow multiple hypotheses. Although the results improve on the Fletcher-Kasturi method, there are difficulties with splitting words apart and over merging multiple words. Other problems include:

- short strings are not detected,
- parallel strings produce false diagonal hypotheses, and
- punctuation signs, dots on the character "i" and other small parts of text are not found.

Finally, they introduce a technique to handle text touching graphics; this can exploit known form information (line locations or configurations) as well as a priori knowledge of line widths. It is assumed that there are no isolated characters, and that some characters of any string do not touch graphics. This allows an island growing approach to be used. Known text elements can be extended, e.g., in rectangular string extension directions. These additions lead to a retrieval rate improvement of about 8% in the five test images.

Techniques from other areas of robotics and image analysis have been applied to the character recognition problem. For example, Wang et al. [127] proposed the use of the Voronoi triangulation method as a means of finding neighboring components which comprise a text string. This is an excellent approach which scales well, too. Henderson and Xu [51] demonstrated a robot navigation method for character recognition. Basically, a virtual robot drives through the line drawing, and the classification of a component depends on the route followed by the robot.

More general character string recognition techniques may also be applied to engineering drawing analysis. Mori [88] gives many useful techniques for character detection and recognition, and we use some similar techniques in the work presented in later chapters. Nishida and Mori [91] propose a model-based split-and-merge method which first merges local segments, then produces more global combinations; they provide a set of strong performance results for handwritten character analysis. Bayesian methods have been demonstrated as well; e.g., see Cho and Kim [23] for

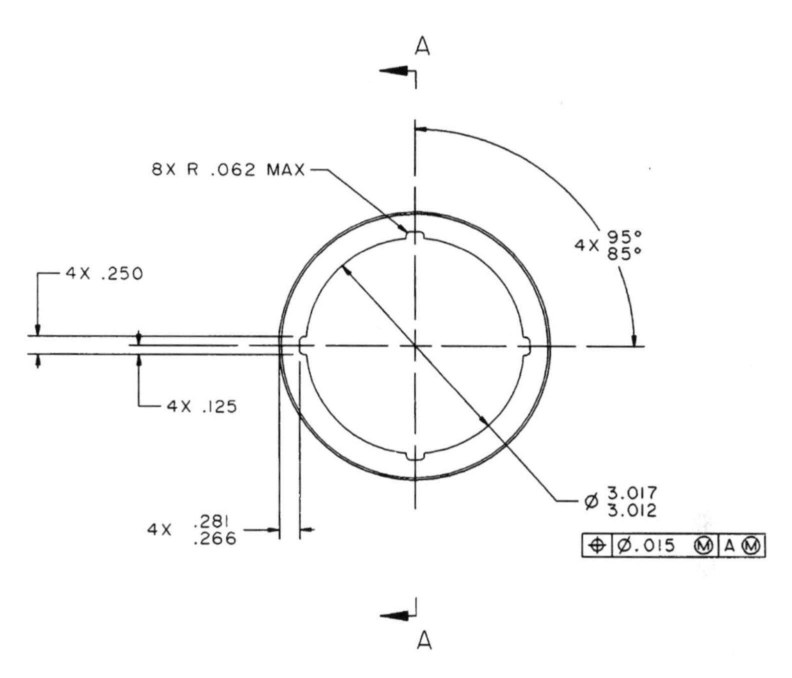

Fig. 1.5 Engineering drawing with dimension sets

an application to handwriting recognition; Barrat and Tabbone [14] showed good results on graphical symbol recognition. Hidden Markov Models (e.g., see [67]) provide another method for character string analysis.

Most engineering drawings follow some standard (or regular) form in terms of the layout of the drawing. Standards are usually established for a group of drawings, and this allows for the application of *form analysis* techniques. General form analysis does not require a priori knowledge of the form structure (e.g., see Wang [125]), but if available, then such knowledge can be exploited (see [48]). Form analysis requires that the image components be classified as boxes, line segment and text components, and then these are analyzed to determine spatial relationships between them. This sort of analysis is demonstrated in a later chapter in terms of extracting the part names and numbers from a set of engineering drawings. Yu [139], Veláquez et al. [124] and Cao et al. [16] all provide methods to separate text characters which touch form or graphics lines.

Another important aspect of engineering drawing is dimension set analysis. Figure 1.5 shows an example of dimension sets in an engineering drawing. Das and Langrana [26] present an early approach for the extraction of dimension sets from engineering drawings. Vectors are input to the system, and dimension set elements are segmented; this includes dimension text, arrowheads, and line elements. These results are then applied to reconstruct the geometry indicated by the drawing. The method was applied to a set of drawings and various types of errors analyzed (e.g., arc and circle locations, dimension values, and coordinates of critical points). The method makes some strong assumptions about the input; i.e., OCR supplies correct information, all parts of a dimension set are recognized (there is no partial recognition of a dimension set), and the vectorized input must be topologically accurate. Dori and Velkovitch [32] present a dimension analysis method based on the ANSI standard. Their text segmentation process includes selection of text segments, growing of regions, connectivity handling, followed by the application of knowledge of the standard. Recognition results are given on a set of hand-drawn documents, and the results are in the 90% recognition range, after a verification step. Su et al. [111] propose a robust method for dimension recognition which proceeds by finding potential dimension frames (i.e., relevant line and text components), and then verifies their spatial relations. They further provide a technique to reconstruct the geometry described in the drawing. Testing on a set of thirty drawings yields about 90% recognition rates. For other work on this topic, also see [29, 65, 128].

A number of systems have been developed for CAD engineering drawing analysis (more general engineering drawing systems have also been proposed, e.g., see [138]). Dori and Liu [34] propose the Machine Drawing Understanding System (MDUS) as a complete scan to 3D geometry reconstruction method. In the version at that time, the processing steps included: vectorization, text recognition, arc segmentation, dimension analysis and hatch line detection. A C++ object oriented system was developed which provided classes of graphical objects. The recognition system followed the hypothesize and test paradigm. They used many of their previously developed algorithms, e.g., the Sparse Pixel Vectorization method, as well as a set of segmentation and classification techniques. A dynamic control mechanism exploits the context of related graphical objects in order to achieve higher classification correctness. Performance results are reported at 94% text, 75% for hatched lines, and 100% for dashed lines (as demonstrated in the 1995 GREC competition).

Ablameyko and Uchida [5] give an overview of current approaches to line drawing entity recognition. This is a very useful summary of the literature in this area. (Note that another useful text on this topic is the book by Ablameyko and Pridmore [2]. The book is complementary to this one in that it covers binarization, connected component analysis, and vectorization in engineering drawings, and also covers the recognition of cartographic objects, and knowledge-based analysis of drawings.) Engineering drawing entities of interest include: contour lines, symmetry axes, hidden contour lines, cross-hatched areas, dimension sets, annotations and circles. Their approach takes advantage of entity features (e.g., line thickness, gaps, structure, and relations between primitive components. An important contribution of the paper is the description of engineering drawing systems extant at the time. Such systems are divided into top-down and bottom-up approaches. In addition to the systems described earlier developed by Dori, Kasturi, Liu, Song, etc., the paper discusses the CELESSTIN [123] and REDRAW [9] systems developed by Tombre. These are rule-based systems based on drawing semantics. The ANON system developed by Joseph and Pridmore [59] uses pre-defined schemata with control rules for bi-directional (top-down and bottom-up) image analysis to identify graphical elements and symbols. Their system provided schemas for various types of lines and curves (solid, dashed), cross hatching, text strings, leader lines, and some cases of dimension sets. Moreover, their control system is specified by an LR(1) grammar which defines strategies by which components may be recognized. Of course, Ablameyko's own work in this area [1, 3, 4] is also reviewed.

A more recent system has been propose by Ondrejcek et al. [93]. In their prototype system called File2Learn, the goal is to discover and record relational information about engineering drawings and 3D CAD models. Various types of information are extracted, including file level information, drawing information contained in the Title Block part of the drawing, and file relationship information. Obtaining meta-data from the file system and the scanned image content allows the creation of ontology elements which can then be used to interrogate the structure and content of the engineering file set. The focus of their work is the recovery of content based information. This is hindered by the low resolution of the scans, the non-conformance to standard layouts of the title block, the variety of fonts as well as hand-written characters, and noise in the scanned images. Although they use commercial software (ABBYY FineReader 9.0) for the OCR aspect of this, there has been some work on this; e.g., see Najman et al. [89]. Najman reports that with their approach to form modeling for title blocks, they achieve about a 70% recovery rate of title blocks, but they say little about the character recognition rates. Ondrejcek reports about an 80% character recognition rate in their work. (For earlier work on title block analysis, see [114].)

More ambitious work has attempted to produce high-level interpretations of engineering drawings. For example, Lu et al. [77] describe a hierarchical knowledge representation method to encode high-level relations between drawing entities, and develop a system to produce such descriptions from scanned drawings. In their case, the application is architectural drawings, however, there are some common issues with mechanical CAD engineering drawing analysis. One aspect they raise is the exploitation of implicit knowledge in the drawings. This includes the multiple views of specific entities, abbreviations, references, inheritance (individual versus system views), symmetries, and dimensions and other annotations. The system has a knowledge representation part, and an interpretation subsystem which exploits the knowledge. A set of descriptors is defined, and then the relational structure is described using an Extended Backus Nauer Form. A set of experiments were run on 271 architectural drawings, and they found that one major issue was that of brittleness. Here that means that the knowledge representation did not allow for imprecise drawings, missing entities or errors in the drawing. However, it does represent an interesting approach. Another attempt at intelligent automatic interpretation of CAD is represented in the work of Prabhu et al. [98]; their method starts with IGES (or DXF) CADD models and discovers features of prismatic (machined) parts.

As for the future direction of this type of research, the field has been widening to include more web related querying. This requires new methods to interact with the user as well as capabilities to analyze documents on the fly. Knowledge representation takes on an indexing flavor, and the classical problems of vectorization, character recognition, and entity segmentation and interpretation become more complicated. Performance analysis is also an important aspect [80]. For a recent survey, see Llados [76].

1.2 Raster Map Analysis

The other major topic of this book is raster map image analysis. Digitized images of maps are called raster maps, and there is a wealth of information contained in such images. However, the extraction and interpretation of the semantic features in raster maps pose significant technical challenges. Figure 1.6 shows a variety of maps and indicates the wide possible range of representations and types of available formats. Figure 1.6a provides a layout of parking lots available in the downtown area of Salt Lake City, while Fig. 1.6b is part of a topographical map of the Salt Lake area. Figure 1.6c shows a historical map of 1849 Texas, and finally, Fig. 1.6d shows the

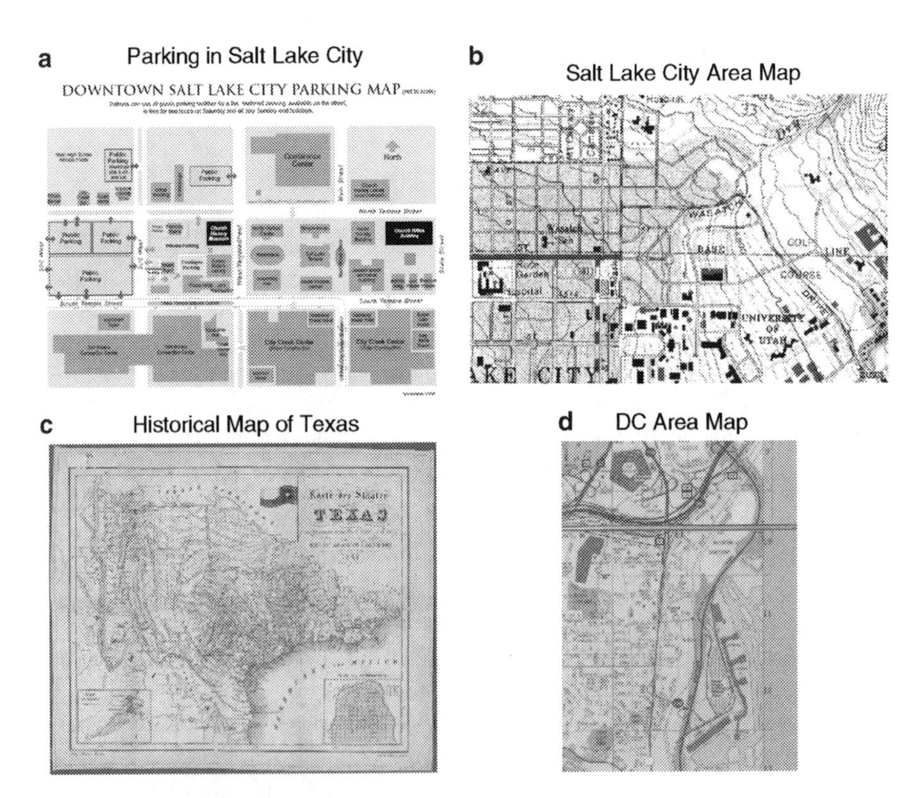

Fig. 1.6 Examples of the wide variety of raster maps: (**a**) A parking map for downtown Salt Lake city. (**b**) Part of a topographic map of the Salt Lake area. (**c**) A historic map of 1849 Texas. (**d**) A map of the DC area

Fig. 1.7 Registration and overlay of extracted semantic map data on an aerial image

DC area with a completely different style of semantic feature expression. Thus, it may be possible that a specific application involves a coherent and homogeneous set of maps (e.g., U.S. Geological Survey maps), but, in general, automatic map analysis methods must be robust enough to handle a wide variety of colors, symbols, textures and layouts.

Chiang [22] has recently described a powerful general approach to high-level raster map analysis and a set of algorithms for the extraction of geospatial information. His specific goals are to exploit roads and road intersections as a basis to separate map layers. Such information can then be registered with other data sources (e.g., aerial images or other maps) to allow further analysis (e.g., change detection, localization, planning, etc.). For example, Fig. 1.7 shows the overlay of extracted map features on an aerial image. Chiang's approach to raster map analysis is shown in Fig. 1.8. His overall goal is the robust segmentation of (1) text and (2) roads, including road intersections and road vectors. Generally speaking, these are the two most useful and informative aspects of a map. Chiang's method proceeds by first decomposing the map into text and road layers. This can be done automatically for high-quality map images, or in a supervised fashion for low-quality images. After this, the text is analyzed in terms of different fonts, sizes and orientations, and semantically identified. The road layer is used to detect road intersections, and this is followed by road vectorization. This approach is not tuned for specific map types, and the results ranged from 99% F-measure on road intersection detection

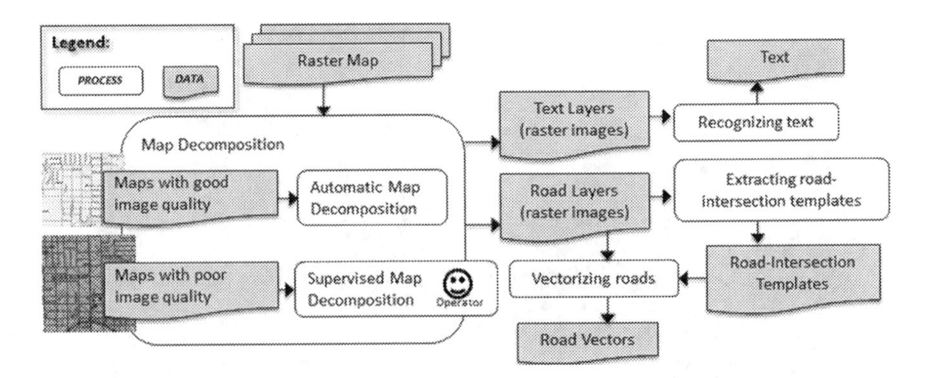

Fig. 1.8 Chiang's raster map analysis process (taken with permission from [22])

in ViaMichelin maps to 69% F-measure on USGS maps (the latter being the main type under study in this book). Also of note is their work on line and character classification (see [132]).

Of course, semantic classes other than roads and text can be found in most maps. Ageenko and Podlasov [6, 95] describe an approach to the basic segmentation of semantic classes in maps. This includes what they term as (1) *Basic*: roads, text and boundaries, (2) *Elevation Lines*: iso-contours, (3) *Waters*: solid bodies and linear, and (4) *Fields*: polygons. They call their method Iterative Semantic Layer Restoration (ISLR), and it is based on the color information and applies mathematical morphological operators to reconstruct the semantic layers after a simple first separation stage. Earlier methods, e.g., that of Khotanzad and Zink [63, 64] studied the problem of basic color recovery based on a scanned image with a wealth of colors; the problem here is that, for example, a USGS map is posited as a small set of colors (eight to twelve or so), and a color scan produces a much larger set of RGB combinations. The approach was to perform eigenvalue line-fitting in the RGB space to recover the best map of scanned colors onto USGS colors. Water and vegetation are segmented based on the color histograms, and linear features are found using valley (in image surface terms) tracking techniques exploiting the A^* algorithm. Evaluation of the results is done in terms of the presentation of processing results. Finally, Henderson et al. [46] discuss machine learning approaches to the automatic classification of a wide range of semantic classes (background, vegetation, roads, water, political boundaries, iso-contours) in raster map images. They describe and compare the results of three unsupervised classification algorithms: (1) k-means, (2) graph theoretic (GT), and (3) expectation maximization (EM). These are applied to USGS raster map images, and performance is measured in terms of the recall and precision as well as the cluster quality on a set of map images for which the ground truth is available. Across the six classes studied there, k-means achieved good clusters and an average of 78% recall and 70% precision; GT clustering achieves good clusters and 83% recall with 74% precision. Finally, EM forms very good clusters and has an average 86% recall and 71% precision. We describe this approach in greater detail in a later chapter.

The recognition of text and symbols in raster map images poses many of the same problems found in engineering drawing analysis. However, the variety and use of text and symbols in raster maps are not as restricted as in engineering drawings. That is, text of different sizes can be overlayed, in different colors, etc., and in any orientation. Early work used mathematical morphology applied to regions of the appropriate color; e.g., Yamada et al. [136, 137] developed a method called the Multi-Angled Parallel method in this way and achieved about 92% accuracy in recognition of elevation text. Kang et al. [61] introduce an interesting method for segmenting distinct text strings using a 3-D graph where the x and y dimensions correspond to the image row and column dimensions, and the z dimension is related to the scale of the text characters:

$$z_i = \frac{Area(c_i)\Gamma}{A_{avg}}$$

where c_i is a connected component, z_i is the z value of a connected component, $Area$ is the area of the enclosing circle of c_i, A_{avg} and

$$\Gamma = \frac{1}{\sqrt{\frac{1}{n}\sum_{i=1}^{n}(Area(c_i) - \frac{1}{n}\sum_{k=1}^{n}Area(c_k))^2}}$$

is the average connected component area. They report good segmentation results with this approach. Li et al. [69, 70] propose a sub-layer separation approach using connected components in each payer, and they report a 98% recognition accuracy. Levanchkine et al. [68] propose to perform text segmentation using various linear combinations of the RGB layers followed by edge detection and gap filling. They recover alphanumeric, point and linear features using what they call *composite images* which are a reduced number of colors to which they apply a hierarchical segmentation. A more standard text analysis is given by Pouderoux et al. [96] which consists of a sequence of steps: (1) segmentation (threshold, erode, dilate), (2) component analysis (based on density and size), (3) string analysis (connecting and merging neighboring connected components), and (4) OCR (using standard methods). This achieves about 96% precision and 92% recall on the tested map images. Roy et al. [100] extract text and symbols by segmenting the map into layers based on color; then for each layer, connected components features and skeleton information is used to identify text and symbols across layers. The method is applied with reasonable success to a set of Russian, Spanish and Bengali maps to give some insight into how it works across alphabets and styles. Chiang and Knoblock [134] give a particularly effective method (exploiting user-specified colors) which also performs a connected component analysis with mathematical morphological operators, and which is robust to character (and string) orientation, overlap, etc.

Given that road information plays a major role in some important applications (e.g., conflation of maps and imagery, road definition for automobile systems, etc.), there are some significant methods to achieve good quality road and road intersection detection. As discussed earlier, Chiang has done much work in this area, and this is well-described in his dissertation [22]; also see [19]. Earlier work by Itonaga et al. [58] used stochastic relaxation to find road labels for pixels in a map image. This step was followed by thinning and then a piecewise linear approximation to get precise road networks. The method was tested by adding noise to perfect images, and for variance of about 0.4, a 90% accuracy rate was observed. This work was done to enhance GPS systems for automobiles. Chiang et al. [21, 133, 135] present a method to obtain road intersections, including their position, connectivity (through roads) and the orientation of roads exiting the intersection. The steps involved included the removal of background pixels, followed by the separation of road lines. Next, road intersections are found using road intersection templates combined with intersection blob detection. This greatly reduces the position error in the location of intersections, as well as the orientation error of intersecting roads. Linton [72] developed a road detection method based on a 3-step process:

1. **Pre-Processing Step**
   - *Histogram Analysis*
   - *Thinning*
   - *Morphological Operations*
2. **Tensor Voting Framework**
   - Fill gaps
   - Smooth curves
   - Produce curve and junction maps
3. **Post-Processing Step**
   - Local Maxima
   - Thinning
   - Thresholding
   - Graph search algorithm

A k-nearest neighbor knowledge-based approach achieves 92% recall and 95% precision on road extraction. We describe this in greater detail later.

Iso-contour segmentation and subsequent terrain surface reconstruction is the final major semantic feature considered here. Of course, the surface reconstruction from contours has been studied for a long time; e.g., Fuchs et al. [38] proposed a method based on finding minimum cost cycles in a directed toroidal graph. Hormann et al. [56] gave a new contour interpolation method which solves the bivariate problem by using a univariate curve Hermite interpolation along gradient directions of the surface (to get smooth transitions across the contours). It is quite efficient and they give a number of examples. A very interesting and effective technique for iso-contour extraction in raster maps is the method of Salvatore et al. [102] which uses a global topology analysis of the layout of contour points by examining the Delaunay triangulation of these points, as well as local geometry which must be satisfied by iso-contours. Shin and Jung [107] give a fast method (about 3X speedup over other methods) for contour-based terrain model reconstruction. Triangle strips are generated using the distance between corresponding vertex pairs on neighboring contours. More recently, Pezeshk and Tutwiler [94] proposed an iso-contour line extraction method using first a quantization of the image, and then contrast-limited adaptive histogram equalization to reduce the effect of noise and semantic feature overlap on contour recognition. Also, see Samet et al. [103].

This concludes our overview of engineering drawing analysis and raster map analysis goals and methods. The following chapters provide greater depth on methods we have developed to solve these problems.

Chapter 2
Segmentation and Vectorization

The segmentation and vectorization of line drawings is well studied[1] (see Chap. 1 for related work), however, no analytic solution has been proposed to date, and therefore, the existing methods are ad hoc and based on heuristics. Most methods also need some amount of tweaking for the specific kind of drawings to be analyzed. We present our approach to these problems here, and this development has been done in the context of the semantic analysis of scanned images of mechanical CAD engineering drawings.

2.1 Segmentation

We assume that a binary image of the drawing is available, and that the line drawing part is the foreground (i.e., pixel value is 1, and usually displays as the color white; note that in this book we will show the foreground pixels as black in figures for better visual clarity). The processing sequence is shown in Fig. 2.1. By segmentation, we mean to divide the skeleton of a connected component into significant pieces that contribute to the intended stroke structure of the component. The input is the image of an individual connected component (e.g., produced by Matlab's *bwlabel*), and the final result is a set of segments from the skeleton. Figure 2.2 shows the set of segments from the image of the digit 3. Each segment consists of a sequence of skeleton pixels (called the *segment path*) that starts and ends with either an *endpoint*, *branchpoint*, or a *virtual point* of the component. In the figure, the endpoints are shown with star shapes, while the single branchpoint is shown as an annulus. Endpoints and branchpoints can be found from simple computations on the neighborhood of a pixel (e.g., see Lam et al. [66]). Matlab's *bwmorph* function can be used to identify these types of points. However, note that *bwmorph* can produce branchpoints that one would not always typically consider

[1] Some parts of this chapter are contributed by Chimiao Xu based on her MS thesis.

T.C. Henderson, *Analysis of Engineering Drawings and Raster Map Images*,
DOI 10.1007/978-1-4419-8167-7_2, © Springer Science+Business Media New York 2014

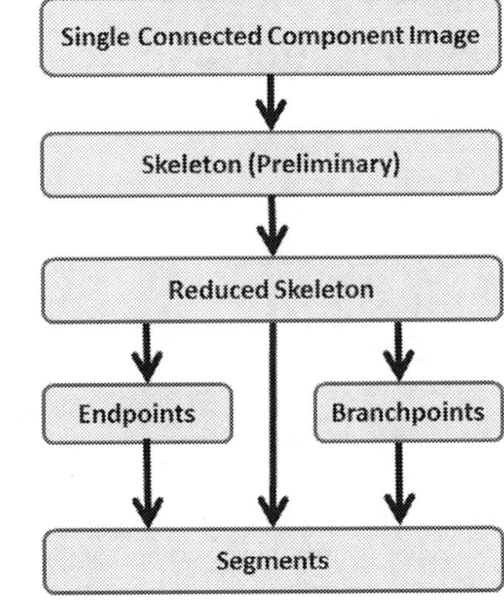

Fig. 2.1 The segmentation process

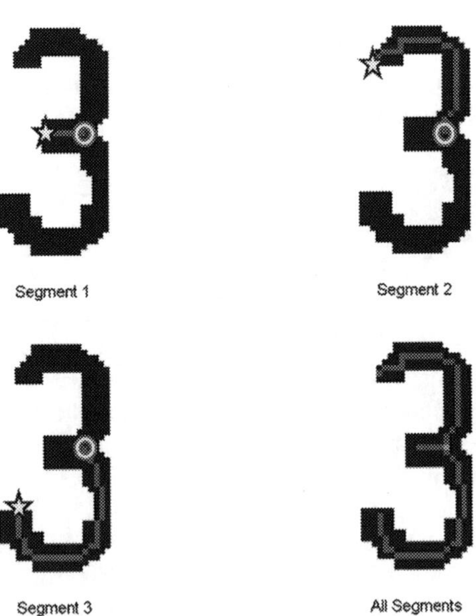

Fig. 2.2 Segments of digit "3"

to be branchpoints. Figure 2.3 shows on the left a set of branchpoints produced by *bwmorph* and on the right our corrected version. In order to eliminate false branchpoints, it suffices to check the following condition on each one: first find the pixels which are three neighbors distant from the branch point—if this set forms three or more distinct connected components, then the branchpoint is valid. *Virtual endpoints* are needed when there is a single cycle (or pixel loop) as will be found in

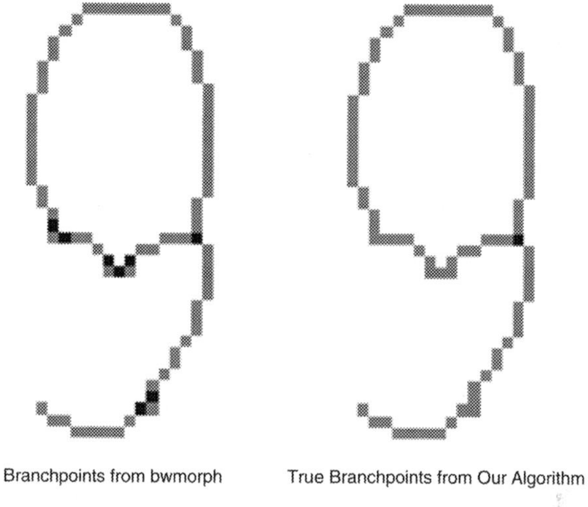

Fig. 2.3 False branchpoints produced by *bwmorph*

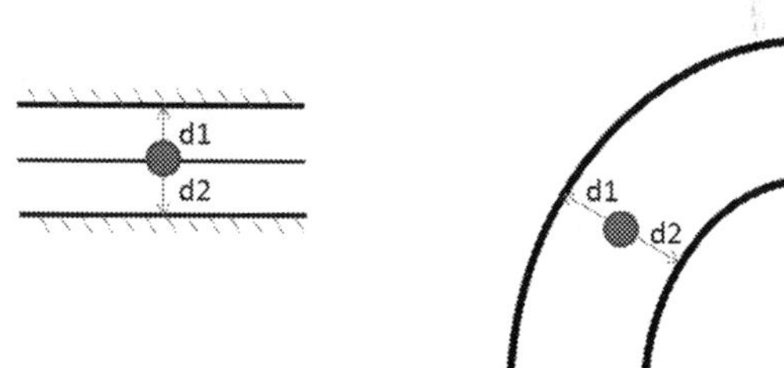

Fig. 2.4 Inside a straight or smoothly bending corridor. The two minimal distance directions to the background are 180 degrees apart

the character "O" and there are no distinguishing points; the *virtual endpoint* is any point of the cycle and is used as both endpoints of the segment path.

Before finding the segment paths, we first eliminate spurious points from the skeleton and produce the *reduced skeleton*. To achieve this, we take advantage of the following observation. Since the line drawings we analyze here are for the most part composed of simple strokes, the vast majority of skeleton pixels will be situated in one of the following contexts:

Context 1: *Inside a straight or smoothly bending corridor* (see Fig. 2.4). In this context, there will be two directions with about the same range value, and these will be the minimum of all range values in a 360 degree scan. These two minima will be about 180 degrees apart. The sum of these two distances is the width of the corridor. Finally, the line defined by the two minimal directions is perpendicular to the direction of the corridor; the skeleton pixels should be running along the middle of the corridor.

Context 2: *Inside a right angle turn in the corridor* (see Fig. 2.5). In a right angle turn, the two minimal range directions are about 90 degrees apart. However, the minimal sum of two opposite directions (180 degrees apart) gives the corridor width and will usually be in the diagonal direction as shown in the figure.

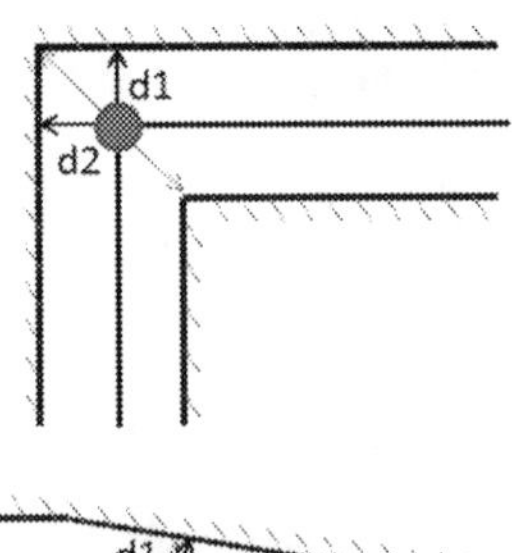

Fig. 2.5 Inside a right angle turn in a corridor

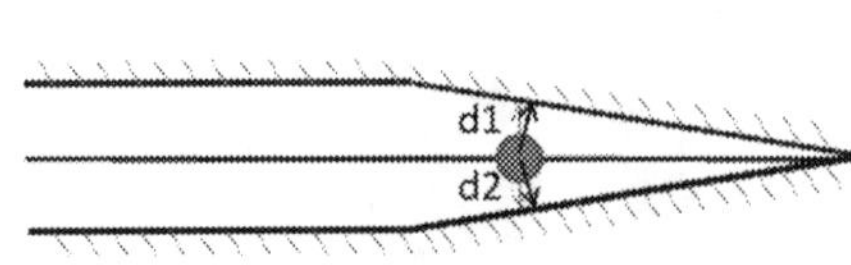

Fig. 2.6 At the end of a corridor

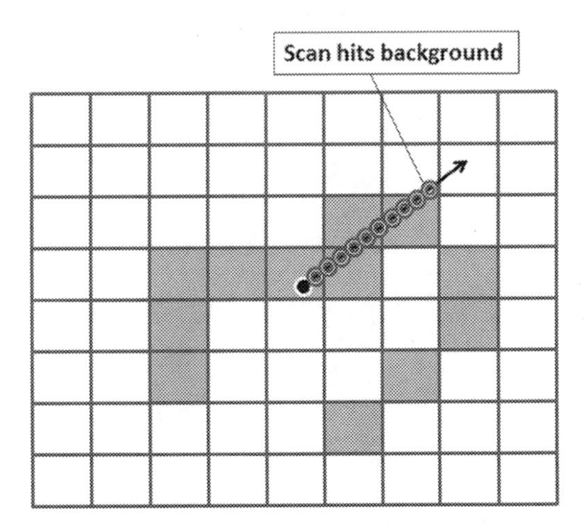

Fig. 2.7 The sub-pixel resolution range scan

Context 3: *The end of a corridor* (see Fig. 2.6). At the end of a corridor, the directions of minimal range to the background will no longer be in opposite directions, but the medial axis will be the bisector of those two directions. The minimal sum of opposite range values will generally be perpendicular to the skeletal axis. This feature distinguishes the right angle turn from the end of corridor and straight or smoothly bending corridor. Note that the end of a corridor may taper together like a pencil point or be more like a the end of a rectangle.

In order to analyze these contexts, we have developed a few image analysis tools. One very useful function is the virtual range finder, *range_scan*. This function determines the distance to the background from a given pixel, starting at zero degrees, rotating in the positive direction (in a right-hand frame; i.e., counterclockwise) in $\delta\theta$ increments until 360 degrees. Our scans are usually in one degree steps. Then in the direction of the scan, we step δx pixels (usually set to 0.001) until a background pixel is reached. Figure 2.7 shows how this works. A sub-pixel length step is made in the scan direction until the step location is in a background pixel. The range map produced by this function is called the *Pseudo-Range Map* (PRM). (Note that it is also possible to obtain a more exact range scan by

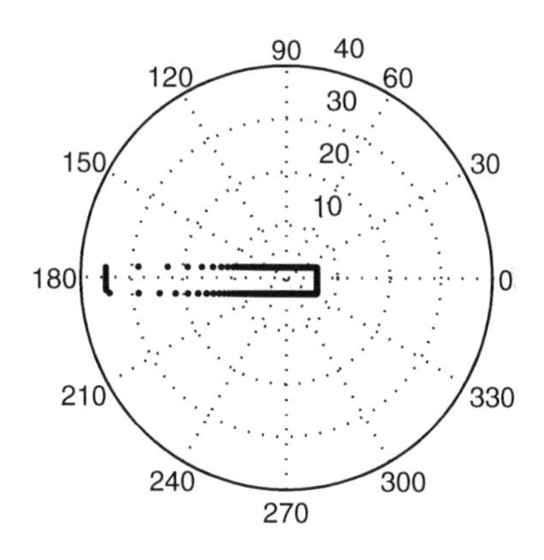

Fig. 2.8 The non-uniform nature of equal-angle sampling in the PRM

intersecting the scan ray with the column and row lines that it crosses, and then give the exact pixel boundary where the change from foreground to background occurs.) The use of robot mapping techniques as a means to understand engineering drawings was the topic of the Masters thesis of Xu [131] who showed that the foreground image could be treated as a floor plan that a small mobile robot navigates in order to segment the components into line segments, endpoints and branchpoints. It was demonstrated there that accurate and robust segmentations could be achieved using this approach which views drawn lines and symbols as hallways and rooms, and the mobile agent is placed *inside* the component and located with sub-pixel accuracy. The Pseudo-Range Map is used by the robot agent to survey the component. (See also Henderson and Xu [51].)

An important issue is whether an equal angle scan should be performed, or a scan that achieves an equal step distance on the boundary. Since the equal angle sample is easier to implement, we have used that, however, in some cases it is better to use the alternative approach since, for example, the center of mass will be highly skewed when most of the scan points are at points near the agent, and far points are sampled less frequently spatially. Figure 2.8 shows a set of scan points and how they are distributed in space (they are plotted as points so that they are more visible). The set of scan points also influences the computation of derived features, such as the

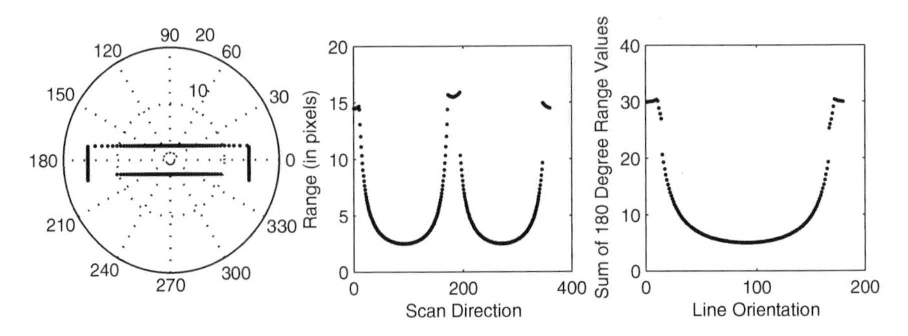

Fig. 2.9 Range scan inside a straight corridor

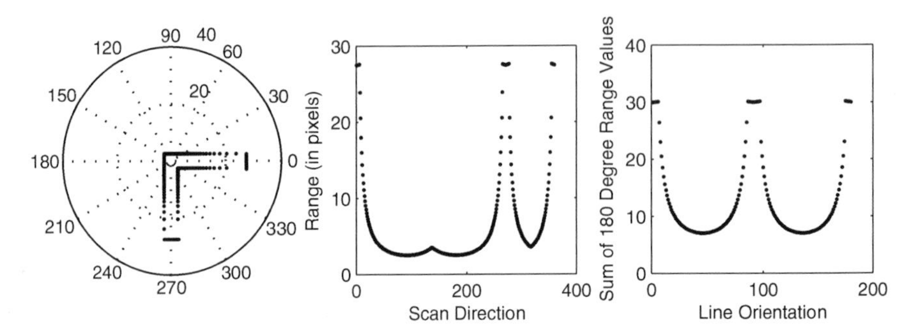

Fig. 2.10 Range scan in a right turn of a corridor

normal at each scan point. As mentioned, another important issue is the resolution of the step in the direction of the scan. The scan distance is found as follows:

- start at the given sub-pixel location
- move δx in the scan direction to a new point x
- stop once x lies in a background pixel.

If δx is too small, then this may be computationally expensive, whereas if it is too large, then background pixels may be skipped over. We have found that $\delta x = 0.001$ provides good pseudo-range scans.

Let's see how this helps determine the context of a skeleton pixel in a corridor. Figure 2.9 shows (on the left) a polar plot of the range to the background for an in corridor pixel; in the center is a plot by direction (0–360 degrees), and on the right is the sum of the two opposite directions of the lines from 0 to 180 degrees. As can be seen the two minimal directions are 90 degrees and 270 degrees (across the corridor) and the minimum sum of opposite directions is at 90 degrees. Figure 2.10 shows the same information for a right turn, and Fig. 2.11 shows the results for the end of a corridor. To produce good skeletons, we proceed as follows. First, *bwmorph* is used to produce a first set of skeleton pixels. Figure 2.12 shows the skeleton produced for the digit four. As can be seen there are a few spurious paths. Figure 2.13 gives the

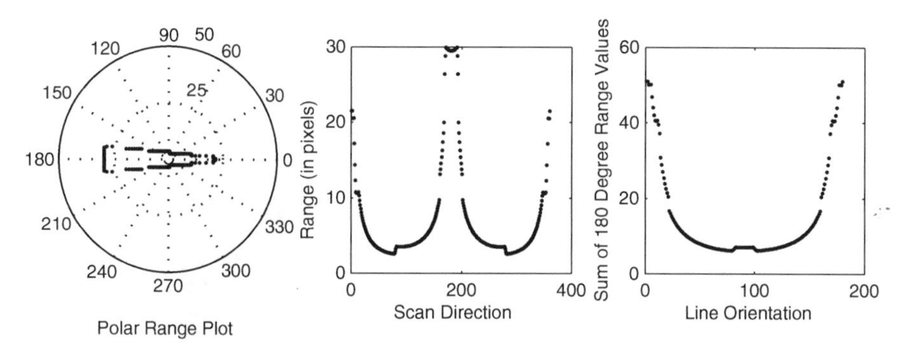

Fig. 2.11 Range scan in the end of a corridor

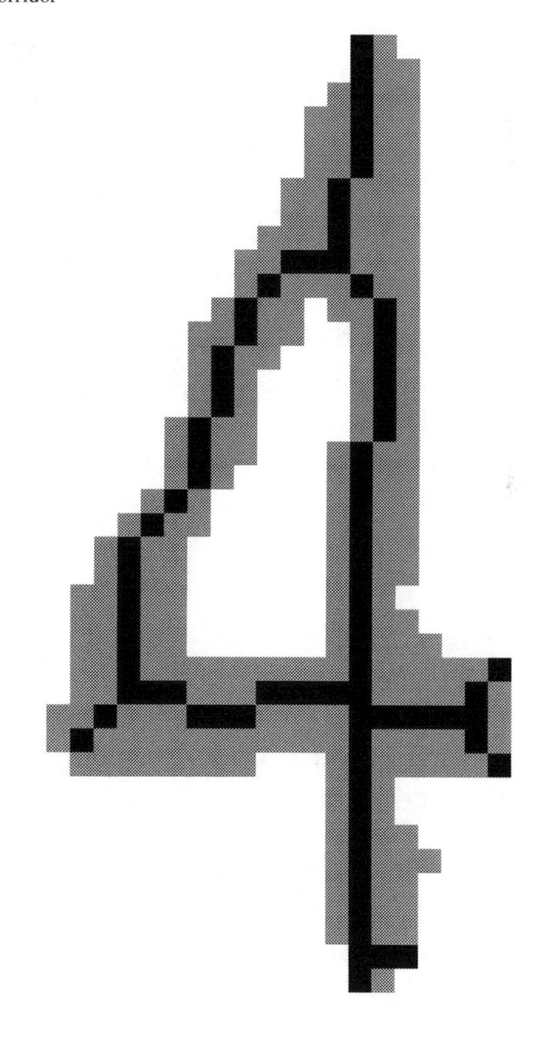

Fig. 2.12 Skeleton produced by *bwmorph*

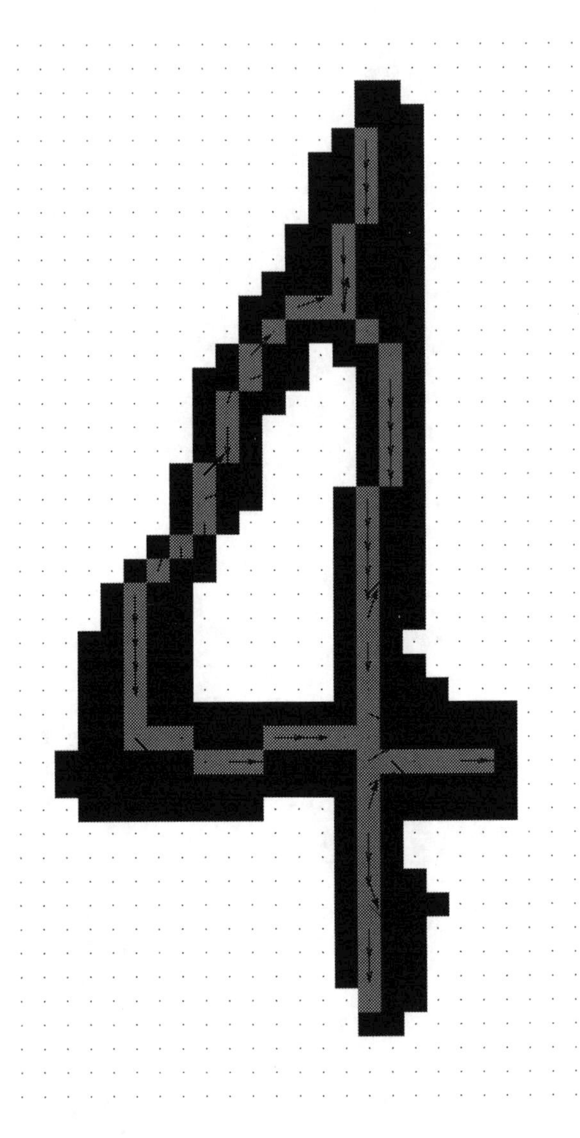

Fig. 2.13 Reduced skeleton produced by pseudo-range map techniques

reduced skeleton produced using the methods described above. Note that the figure also shows the estimate of the corridor direction produced for each pixel. Figure 2.14 shows the skeletonization result on some images of the ten digits, while Fig. 2.15 shows the same for the first ten letters of the alphabet. Figure 2.16 shows the results on the arrow part of a dimension set.

Xu studied two important aspects of the engineering drawing analysis problem: point feature analysis and linear feature analysis. The PRM was used pixelwise to identify:

1. *endpoints*: the terminal part of a line segment.

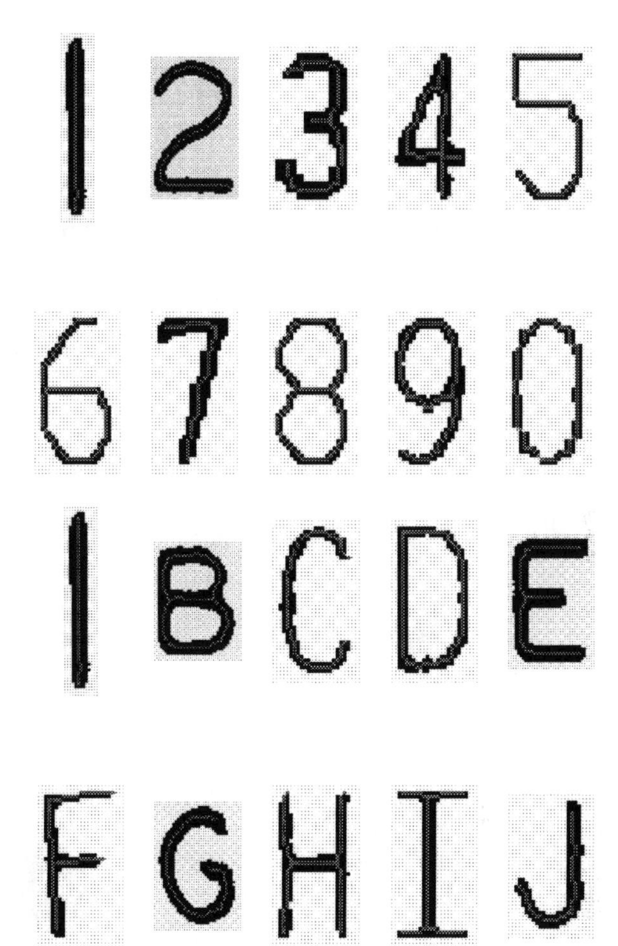

Fig. 2.14 Skeleton produced for digits

Fig. 2.15 Skeleton produced for letters

2. *interior corridor points*: two directions of travel are possible, but not a corner.
3. *corner points*: two directions of travel possible, but at significantly different angles.
4. *multibranch points*: more than two directions of travel possible.

Methods were developed to identify these four types of point features, and a decision tree method was used to classify pixels. A set of attributes of the pixel and its PRM were selected, and at each nonterminal node of the tree an attribute value is used to pick a branch to follow, and each leaf node provides a point feature decision (i.e., one of the four possible types). An information theoretic approach was used to inform the construction of the decision tree (see [101]). Attributes used in building the decision tree included:

- **Attributes 1:7**: Hu invariant moments [57] of the PRM.
- **Attribute 8**: Area of the PRM region.

Fig. 2.16 Skeleton produced for arrow of dimension set

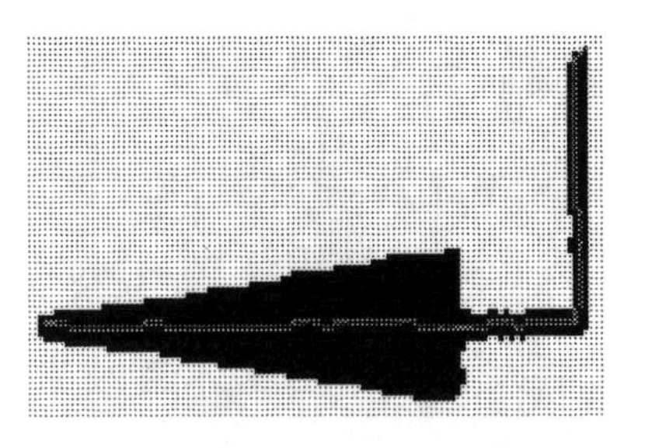

Skeleton for an Arrow

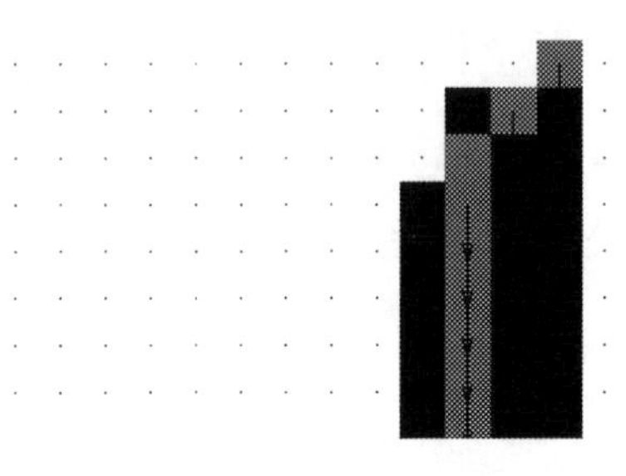

Detail of Upper Part of Arrow

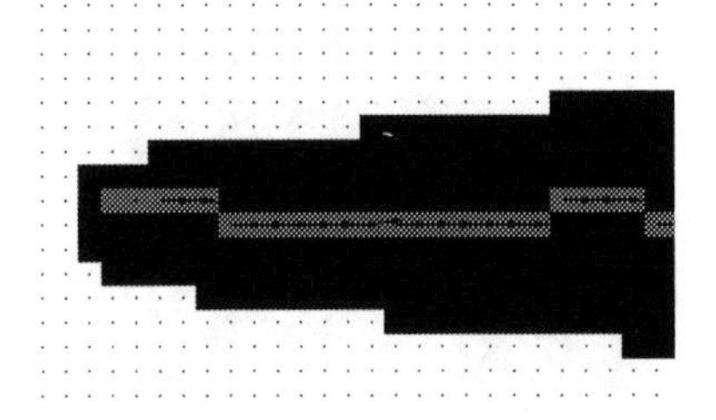

Detail of Point Area of Arrow

- **Attribute 9**: Perimeter of the PRM region.
- **Attribute 10**: Sum of ranges of PRM.
- **Attribute 11**: Total distance of the points in the PRM perimeter to the point that is used to compute the PRM.
- **Attribute 12**: Sum of the absolute distance of the rows and columns of the PRM region points to the row and column of the point used to compute the PRM, divided by the total number of points in the PRM region.
- **Attribute 13**: Number of branches on the PRM.

Table 2.1 The *recall* and *precision* results for the non-PRM, PRM and decision tree methods

	F_end	F_corr	F_corn	F_bra	
Non-PRM	91.94	99.26	64.32	83.10	
PRM	98.59	98.73	99.93	98.96	*Recall*
D-Tree	98.92	86.38	73.41	93.80	
Non-PRM	45.62	100.00	43.12	40.13	
PRM	52.14	100.00	41.10	49.98	*Precision*
D-Tree	33.38	100.00	34.16	35.65	

The attributes with the most discriminatory power for each feature type were: (1) endpoints: Attribute 11, (2) corridors: Attribute 2, (3) corners: Attribute 12, and (4) branches: Attribute 11.

The overall system performance was evaluated in terms of the quality and computational complexity demonstrated over various image datasets. Since noise occurs in the images, there are extraneous as well as missing objects. In order to objectively measure how well the proposed system analyzed digitized engineering drawings, we compared the results over a dataset for which ground truth was available. The following steps were taken:

- A testbed benchmark set of five images and ground truth were established.
- A non-PRM classifier was run on the images.
- A hand-made PRM classifier was run on the images.
- A decision tree classifier was developed using training data and run on the dataset.

To measure the performance, we use *recall* to mean the ratio of correct features found to the total number of relevant features, and *precision* to mean the ratio of the number of correct features found to the total number of features found. Table 2.1 gives the results of the performance analysis.Although the static PRM feature classification method works well as far as *recall* is concerned, good precision can only be achieved by applying a post-processing step.

Xu also developed methods for linear feature extraction. As described in Chap. 1, much work has been done on this topic, and our method can be viewed as an extension of the *Zig-Zag* method of Dori [30]. The mobile robot mapping approach exploits the Generalized Voronoi Diagram (GVD) [12]. After constructing the GVD, linear segments are found by using the differential GVD curvature.

In the free space of a plane with a set of obstacles, the Voronoi diagram (VD) [24] is defined as a collection of line segments or convex polygons that partition free space into n cells so that each cell contains one obstacle. Any point on the edge of two neighboring cells is equidistant to the two particular obstacles in the two neighboring cells. When the obstacles extend to any object, the VD is called the generalized Voronoi diagram (GVD) for these objects. In the robotics path planning community, the GVD is used as a roadmap through a planar set of obstacles. The planar space is divided into *free space*, or a set of points through which the robot may pass, and the *obstacle points* through which the robot may not pass. Roadmaps provide a structure completely describing the topology of the workspace, including: accessibility, connectivity and departability. Robot motion planning consists of:

1. Find a way to get to the GVD.
2. Find a path along the GVD to the neighborhood of the goal point.
3. Find a way from the GVD to the goal point.

The GVD is more formally considered a deformation retract, defined in terms of a continuous map from the free space of the GVD. In our problem domain, we are only interested in the GVD points found in the corridors (i.e., curvilinear segments); at segment ends, we want the points to go straight to the end of the line segment rather than splitting and going to the corner points as occurs with the medial axis transform.

The modified GVD (MGVD) points, or perhaps more appropriately, the center line points, are characterized by being midway between two scan points that are 180 degrees separated in the range scan, are a minimally distant pair of 180 degree separated points, and have inward pointing normals. The fundamental idea of the segmentation algorithm introduced by Xu is based on the fact that every foreground pixel must belong to a linear segment or multiple segments if the pixel is either a corner or branch point. The PRMs of most points in a given linear segment share similar features captured by the PRMs. Such points are usually interior corridor points. The algorithm takes the following steps to extract a MGVD path as a linear segment.

1. Find an unexpanded interior corridor point **P**.
2. Compute **P**'s PRM and the longest forward direction **FDIR**.
3. Compute a MGVD path along **FDIR** and its opposite direction **BDIR** until it hits the background pixel boundary.

The set of MGVD nearest neighbor points form a linear path through the segment. Hence the linear segment to which **P** belongs has been extracted. By iterative application of the above steps on a connected component foreground object, the algorithm extracts all linear segments of the connected component. Performance measures for the MGVD extraction method were developed and tested on several engineering drawing images. Correspondence to ground truth segments is made based on:

1. The distance between segment endpoints **gpt1** and **pt1**, and distance between the other segment endpoints **gpt2** and **pt2** must both be less than the segment's width.
2. The absolute difference between **GDIR** and **DIR** must be less than 5 degrees out of 360.

- **gpt1** and **gpt2** are the two endpoints of a ground truth segment, and **pt1** and **pt2** are the two endpoints of an extracted segment.
- **GDIR** is the ground truth segment's direction, and **DIR** is the extracted segment's direction.

The MGVD method was compared to a Hough-based method on eight test images, and the average correctness result for MGVD was 99.63% whereas the Hough method was at 82.98%.

In summary, Xu's work explored the use of robot navigation and mapping techniques in the analysis of engineering drawings. Contributions include:

- *Definition and Analysis of the Pseudo Range Map (PSM)*: For each foreground point (to sub-pixel accuracy), the distance to the background is computed for a selected set of angles and to sub-pixel accuracy. The PRM has been shown to provide a robust and stable basis for mapping analysis.
- *Point Feature Analysis*: Given the PRM at a specific location we have demonstrated that the shape of the PRM permits classification of the location into one of the categories: endpoint, corridor point, corner point or branch point. The correct classification rates are very high compared to conventional techniques.
- *Linear Feature Analysis*: Given the ability to move an agent in the foreground, we have demonstrated the extraction of an approximation to the Modified Generalized Voronoi Diagram (MGVD) and can exploit the MGVD to segment the drawing into 1D curves and straight line segments.

2.2 Vectorization

Once a set of segments is identified, the next step in the process is to produce a higher-level approximation to the segments; recall that a segment consists of a set of pixels running from an endpoint to an endpoint, or an endpoint to a branchpoint, a branchpoint to a branchpoint, or a virtual endpoint to itself. For the application at hand, we use a straightforward line fitting algorithm which tracks from one segment endpoint to the other taking in new pixels until a fitting threshold is exceeded. After a piecewise polyline is produced, the endpoints are adjusted for a better fit for the two vectors to either side.

Figure 2.17 shows the vectors produced for the digits.

2.3 Connected Component Information

Although the segments and vectors are the core information about each connected component, it is useful to put together more information about the component which will make further analysis easier. This information includes:

- *image*: original image
- *branchpoint image*: specific branchpoint pixels
- *skeleton*: center line points through component
- *segments*: the segments as described above; for each segment:

Fig. 2.17 Vectors produced for digits

- *path*: row, col array from one endpoint to the other
- *num_rows*: number of rows in the original image
- *num_cols*: number of columns in the original image
- *endpoints*: endpoints of the segment; these are actual termination points (i.e., not branchpoints or virtual points)
- *branchpoints*: branchpoints of the segment

- *vectors*: the vectors as described above; for each vector:
  - *points*: row, col array of points in the vector
  - *num_pts*: number of points in the vector
  - *index 1*: segment path index for first point of vector
  - *index 2*: segment path index for last point of vector
  - *error*: error for vector fit to segment
  - *endpoint 1*: row, col of first point
  - *endpoint 2*: row, col of second point
  - *theta*: orientation of vector
  - *num_rows*: number of rows in original image
  - *num_cols*: number of cols in original image
- *segment neighbors*: s by s array of segment neighbors
- *vector endpoint info*: for each endpoint, first and second element are row, col of endpoint, and third element is the number of vector neighbors
- *vector neighbors*: $2v \times 2v$ array of vector endpoint neighbors
- *cycles*: cycles found in the component
  - *paths*: list of vector endpoint sequence that defines the cycle
- *vector tree*: tree of vector connectivity; each node has:
  - *parent*: index of parent node

  - *index*: vector endpoint index
  - *children*: indexes of children nodes

- *paths*: endpoint to endpoint paths (starting vector is indexes 1 and 2)

  - *path*: sequence of vector endpoint indexes from one endpoint to the other

These items are useful when trying to recognize characters, arrowheads, graphics, etc. For example, the digits “0”, “4”, “6”, and “9” have one cycle (unless it is broken due to noise or touching another component); moreover, the digit “8” has three cycles: the upper and lower two circles as well as one around the outer periphery. The path information allows one to more easily find, e.g., the correct one path through the segments which are produced for the character “2” (alternative paths may be created by spurs and noise). The exploitation of this information will be further elucidated in following chapters.

Chapter 3
Text and Graphics Analysis in Engineering Drawings

The meaning of an engineering drawing is expressed through text and graphics and the relations between them. Chapter 1 provided a detailed summary of the major approaches to their segmentation, and here we describe our own contributions on some specific applications. The goal is the fully automatic segmentation of text and graphics in an engineering drawing image, as well as its interpretation; that is, characters represented as pixels must be interpreted as to which specific character they represent. Of course, this is made difficult in that most engineering drawings use a variety of fonts, sizes, and orientations for characters—indeed, some are even hand-written. In addition, character segmentation is generally only part of a larger process: for example, dimension set analysis. Since the names and numbers extracted by the system are quite significant for manufacturing purposes, say in a reverse engineering application, then more likely than not, hypotheses put forward by the image analysis system will need to be corroborated by a human.

Consider the following problem scenario:

> **Problem 1**: A dataset consisting of a large number of paper drawings (perhaps several hundred) is the only extant record of the CAD development of a vehicle still in use and which needs to be modified. The drawings have been scanned, and a collection of digital images created—one for each drawing. The goal is to create a database for the drawings which contains meta-data about each drawing (type of drawing, name of part, part number, etc.), as well as the contents of the drawing (e.g., graphics, dimension sets, etc.).

To solve this problem, we break it up into several sub-tasks. Generally, there are several drawing types, including general arrangement drawings, arrangement drawings, assembly drawings, detail drawings, and fabrication drawings. The first task is to determine which types of drawings exist in the dataset and their formats. Sometimes, other types of information are provided; for example, in a study we performed, files like that shown in Fig. 3.1 were included; these are best interpreted using standard commercial off-the-shelf OCR software. More typical is the detail drawing shown in Fig. 3.2. The solution requires the identification of the *form* (i.e., the specific layout of lines and text structure of the key), and this, in turn, is based on the extraction of text and straight lines in specific areas of the drawing. For example,

T.C. Henderson, *Analysis of Engineering Drawings and Raster Map Images*,

DOI 10.1007/978-1-4419-8167-7_3,

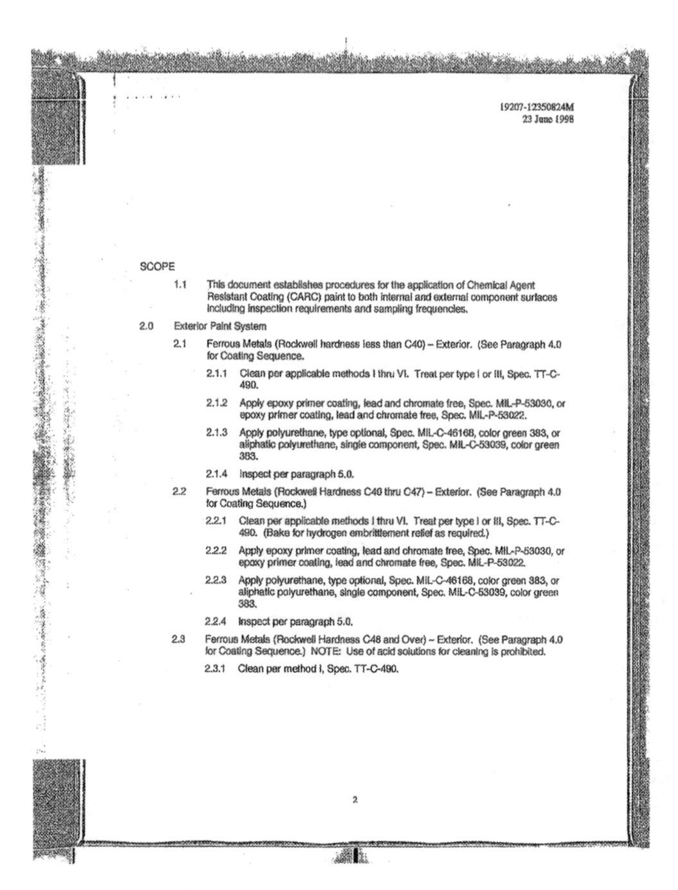

19207-12350824M
23 June 1998

SCOPE

1.1 This document establishes procedures for the application of Chemical Agent Resistant Coating (CARC) paint to both internal and external component surfaces including inspection requirements and sampling frequencies.

2.0 Exterior Paint System

2.1 Ferrous Metals (Rockwell hardness less than C40) – Exterior. (See Paragraph 4.0 for Coating Sequence.

2.1.1 Clean per applicable methods I thru VI. Treat per type I or III, Spec. TT-C-490.

2.1.2 Apply epoxy primer coating, lead and chromate free, Spec. MIL-P-53030, or epoxy primer coating, lead and chromate free, Spec. MIL-P-53022.

2.1.3 Apply polyurethane, type optional, Spec. MIL-C-46168, color green 383, or aliphatic polyurethane, single component, Spec. MIL-C-53039, color green 383.

2.1.4 Inspect per paragraph 5.0.

2.2 Ferrous Metals (Rockwell Hardness C40 thru C47) – Exterior. (See Paragraph 4.0 for Coating Sequence.)

2.2.1 Clean per applicable methods I thru VI. Treat per type I or III, Spec. TT-C-490. (Bake for hydrogen embrittlement relief as required.)

2.2.2 Apply epoxy primer coating, lead and chromate free, Spec. MIL-P-53030, or epoxy primer coating, lead and chromate free, Spec. MIL-P-53022.

2.2.3 Apply polyurethane, type optional, Spec. MIL-C-46168, color green 383, or aliphatic polyurethane, single component, Spec. MIL-C-53039, color green 383.

2.2.4 Inspect per paragraph 5.0.

2.3 Ferrous Metals (Rockwell Hardness C48 and Over) – Exterior. (See Paragraph 4.0 for Coating Sequence.) NOTE: Use of acid solutions for cleaning is prohibited.

2.3.1 Clean per method I, Spec. TT-C-490.

2

Fig. 3.1 A non-standard file in the dataset

in Fig. 3.2, the name of the drawing (the part name, in this case) is "RETAINER, PACKING" and the Part No. is 8737863. To find the legend with the name and part number requires the extraction of text and straight lines with specific relations holding between them.

3.1 Text/Graphics Segmentation

The text/graphics segmentation process is shown in Fig. 3.3, and the various steps can be performed automatically or on the basis of user input. For example, if the input consists of gray-level images, then a threshold must either be specified or discovered using histogram analysis combined with the fact that the foreground comprises around 20% of the image pixels.

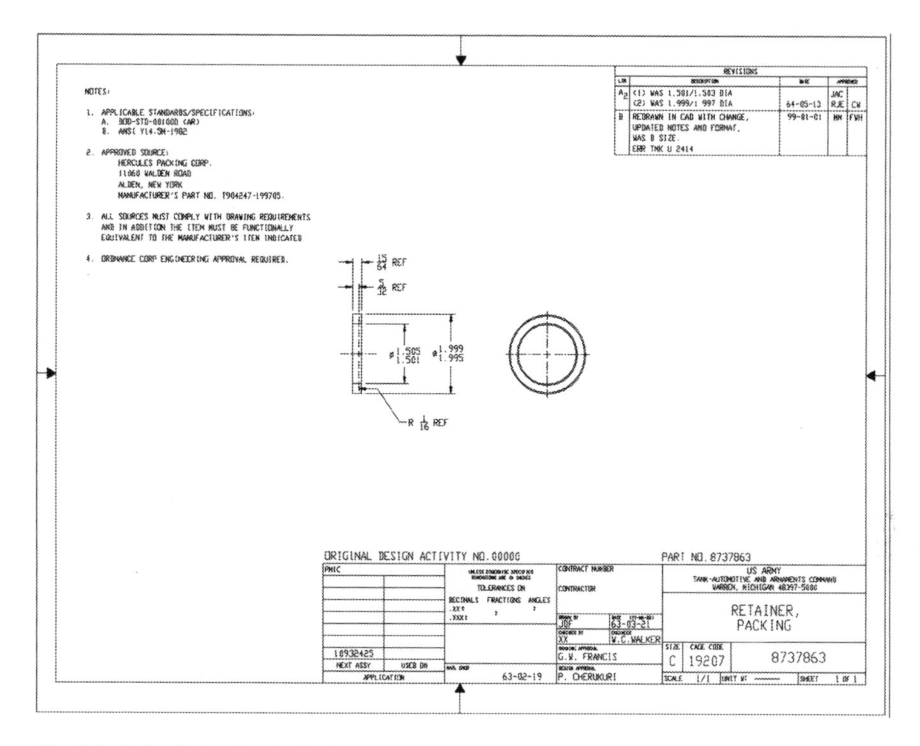

Fig. 3.2 A detail drawing in the dataset

3.1.1 Form Analysis

Most engineering drawings follow some standard (or regular) form in terms of the layout of the drawing. Standards are usually established for a group of drawings, and this allows for the application of *form analysis* techniques. General form analysis does not require a priori knowledge of the form structure (e.g., see Wang [125]), but if available, then such knowledge can be exploited (see [48]). Form analysis requires that the image components be classified as boxes, line segment and text components, and then these are analyzed to determine spatial relationships between them. Liang [71] proposes a technique for the analysis, extraction and classification of a document layout structure as a set of bounding boxes of the constituent connected components in the document. They extract a document layout structure using a bottom-up approach while we use a top-down approach to extract boxes and then the connected component structures lying in it. Also, see [126]; they propose a framework to identify and classify a given form as an explicit, semi-explicit or implicit style form depending on the presence of lines bounding the elements in the form. Also see the discussion in Chap. 1 of the work done by Ondrejcek et al. [93].

Chhabra [20] describes a very similar system in function to that described here. However, there are several significant differences. His work addressed telephone

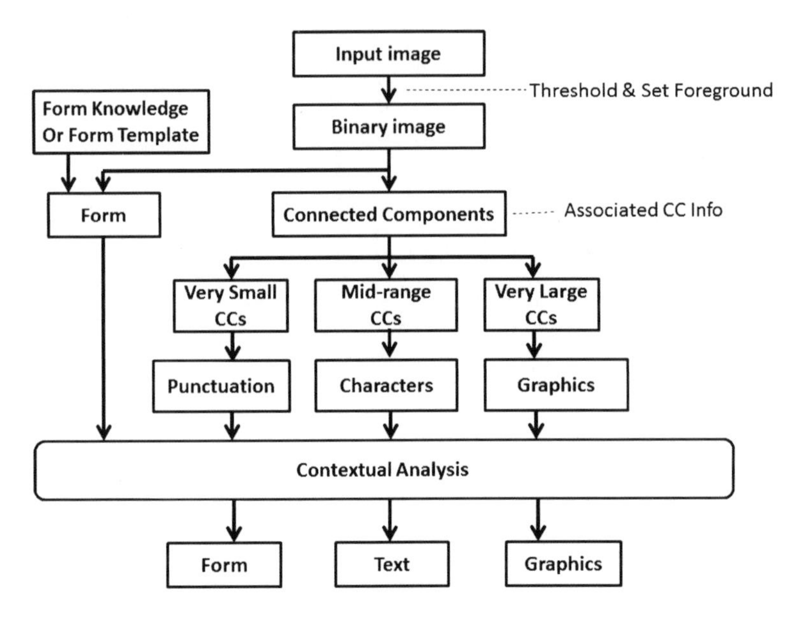

Fig. 3.3 Basic text/graphics segmentation

company forms and used a great deal of prior knowledge (pre-defined grammars for the table entries). More preprocessing was performed (deskewing), complex features are computed on the characters (we simply use a normalized version of the character), and the neural net had to be trained on tens of thousands of sample characters (we use just nine samples per character). Finally, no performance results are stated for their work.

The relations between logical parts of a layout provide a way to describe a form. Several form identification methods have been proposed in the past, including that of Diana et al. [27] which was used in the analysis of documents from the French social services department (les Allocations Familiales). A bottom-up approach that segments text, lines, and boxes into components, recovers the relations between text and boxes, and uses the known line structure to separate text from touching lines was proposed by Wang and Shirai [125]. Good results on model-based analysis and understanding of checks were demonstrated by Ha and Bunke [41]. Arai and Odaka [10] showed results on fifty different types of forms with a box extraction technique based on a background region analysis. Finally, semantic information extraction from forms as well as efficient storage was described in work by Tang and Lin [116].

For our analysis, a form model must be given (or perhaps learned) and is expressed in terms of spatial relation constraints which define the structural model. If boxes (i.e., rectangles) are square to the image frame, then the sides can be numbered from 1 to 4, starting at the top and going clockwise. An example form structure is shown in Fig. 3.4. Then a simple grammar for this layout could be:

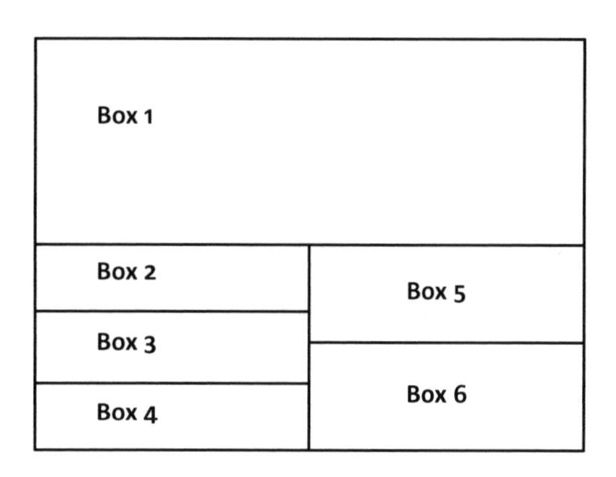

Fig. 3.4 An example form layout

```
Rule 1: form := A + B + C
        [side(A,3) intersects side(B,1)]
        [side(A,3) intersects side(C,1)]
        [side(B,2) intersects side(C,4)]

Rule 2: A := Box1

Rule 3: B := Box2 + Box3 + Box4
        [side(Box2,3) = side(box3,1)]
        [side(Box3,3) = side(box4,1)]

Rule 4: C := Box5 + Box6
        [side(Box5,3) = side(box6,1)]
```

The terminal symbols are simple rectangles. The non-terminal symbol A represents the top box; B represents the lower-left three boxes, and C represents the lower-right two boxes. The square brackets in each rule define the constraints as predicates on the rectangles. A more detailed discussion of results will be given later when the complete document analysis system is described.

3.1.2 Connected Component Analysis

There are many algorithms to segment and label connected components in binary images (e.g., Matlab provides *bwlabel*), where a connected component (CC) consists of a set of connected pixels (i.e., there is a path through the set between any two pixels in the set). Every pixel in a connected component is given the same value as the connected component index from 1 to n, where there are n CCs. Depending on the quality of the input image, it may be necessary to pre-process the image to reduce the effects of noise, blurring, etc. This must be done carefully since low bandwidth filtering may cause distinct components to merge, and high-bandwidth filtering may disconnect components.

Thus, connected components may include multiple semantic objects (e.g., a character and a line segment) which will require a more sophisticated analysis. The cleaner the initial segmentation in to CCs, the easier the text/graphics segmentation will be. Another major difference between standard text document analysis and engineering drawing text segmentation is the relatively small number of characters in the latter. Therefore, it is necessary to be able to recognize isolated characters at any orientation and scale. We assume for the present that characters occur as distinct CCs.

The next step is to produce a set of basic comprehensive information about each connected component. The set of information consists of the following:

- *character semantics*: which character is represented by the CC.
- *character image*: the binary input of the character; each CC is represented in an image so that there are foreground pixels in the first and last rows and columns.
- *bounding box*: the corners of the smallest bounding sub-image in the original image.
- PCA frame: the PCA frame is found for each set of foreground pixels.
- *CC skeleton*: this should be a medial axis like set of pixels.
- *branch points*: points in the skeleton where three paths diverge.
- *endpoints*: extreme points on a segment of the skeleton.
- *segments*: all points in the skeleton are either *branch points*, *endpoints* or *regular points* (i.e., regular is neither end nor branch point); a segment is then either (1) a linear sequence of skeleton points with a branch or endpoint at each end and regular points in between or (2) a cycle consisting of only regular points.
- *endpoint to endpoint paths*: the set of all linear paths running from one endpoint to another.
- *cycles*: the set of all unique cycles in the skeleton.
- *vectorization*: a vectorization of the skeleton.

This set of information informs the character analysis process. Let's consider some examples of this. Figure 3.5 shows a set of eight digits and upper case letters taken from an engineering drawing, while Fig. 3.6 shows the skeletons, endpoints and branch points for these CCs.

3.1.3 Character Analysis

Characters and symbols are human artifacts and should be variations on some ideal set of intended strokes. However, such variations can make automatic recognition difficult. For example, the letter "A" is shown in Fig. 3.7 in a number of fonts. On the one hand, character recognition can be based on a principled or rule-based approach, or on the other using a data-centric machine learning approach; we will

Fig. 3.5 Set of connected components from digits and letters

Fig. 3.6 Skeletons, branch and endpoints for set of connected components

Fig. 3.7 Variety of different representations of letter "A"

demonstrate examples of both. It is also necessary to decide whether a single class is to be assigned to each CC, or a probability density function across all possibilities.

It may be possible to filter the set of CCs to be considered for text analysis. A histogram of the CC sizes allows them to be split into categories: (1) very small, (2) mid-range, and (3) very large. The very small CCs are usually either noise, punctuation, parts of a letter (e.g., the dot on an "i" or "j" or a disconnected piece of a character), or annotation glyphs (e.g., the dots in a dotted line). The very large CCs are generally either graphics or part of the form for the engineering drawing (i.e., the set of lines and boxes which separate out meta-information about the drawing). Graphics are predominantly linear features (straight or curved), and forms consist of

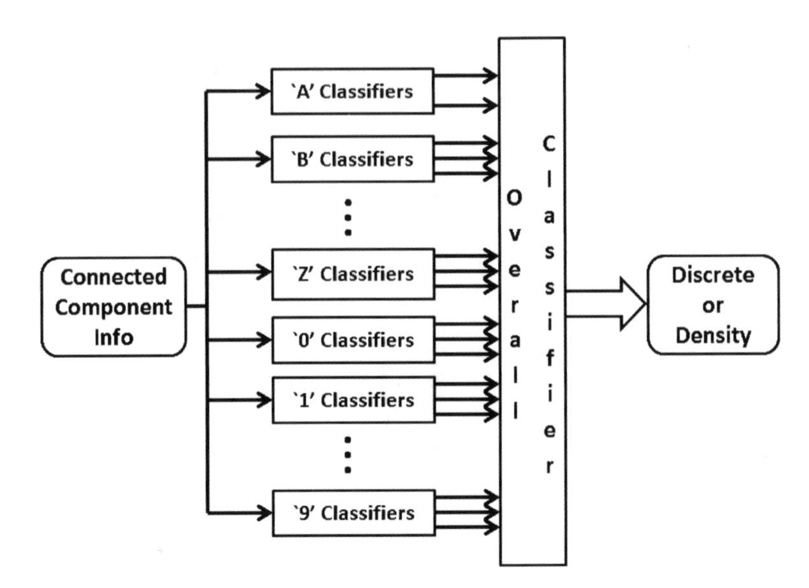

Fig. 3.8 Character analysis layout

rectangles aligned with the frame of the drawing. Characters and symbols are mostly found in the mid-range size CCs. An important and distinct class is the set of simple straight line segments (e.g., the digit "1", the letter "l", dashes "-", slashes "/", etc.). These are typically three times longer in one dimension than the other. We call these *linear* CCs. When segmenting text, their bounding boxes provide information about neighboring characters in that they should mainly differ by a translation (of course, the height and width may vary some, but most lower box sides will line up along the bottom of the text). In our discussion of character segmentation here, we consider the digits $\{0, 1, 2, 3, 4, 5, 6, 7, 8, 9\}$ and the upper case letters "A" to "Z". There are many character classification techniques as described previously, and we compare some simple approaches which work well on the meta-data extraction scenario posed at the start of the chapter.

A Rule-based Method We describe first a method which uses a set of decision functions all applied to each CC, and each of which produces a value between 0 and 1 as to the likelihood of the CC being that specific character. If there are ties and a unique decision is desired, then some sort of tie-breaking function must be used. Alternatively, the results of each classifier may all be sent to a function which uses that plus contextual information to make a final assessment. Figure 3.8 shows the analysis flow. Note that the result may be either a discrete class or a probability density function over all possibilities.

Consider now how the features of a character, e.g., the endpoints, branch points, cycles, etc., can be used to construct a recognizer. The digit '2' has the following features:

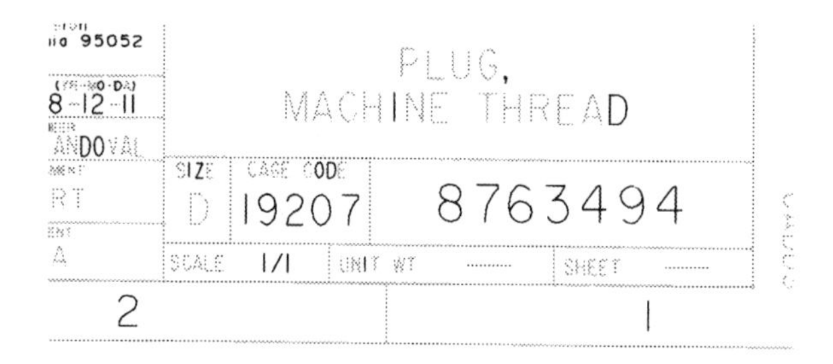

Fig. 3.9 Results of rule-based digit recognition

- *cycles*: none
- *branch points*: none
- *endpoints*: two (of course, due to noise, etc. there may be extraneous endpoints, and thus, branch points as well. The recognition process must be robust to these possibilities)
- *special features*:
  - Although there may be several endpoints, a '2' will be characterized by the fact that the longest endpoint to endpoint path will be from the upper left one to the lower right one. Therefore, first find the longest such path (the CC information contains the set of all such paths).
  - The two endpoints of the longest path must satisfy the requirement that one be in the upper left quadrant of the CC image, and the other be at the lower right part.
  - There must be a part of the character that covers most of the columns along the bottom part of the image.
  - The longest path should cover most of the pixels in the skeleton of the character.

Matlab functions have been written for the digits, and Fig. 3.9 shows the digits recognized in part of an engineering drawing. As can be seen, the method is very effective when the style of the digits is restricted. There are some errors; for example, the letter "D" is labeled "0". If letter recognizers were implemented as well, then the overall classifier would need to decide which of the two labels would be more likely, or put forward multiple hypotheses. As we shall see later, the latter approach allows a more robust analysis in keeping all reasonable hypotheses and use higher-level context to make a final decision. This was proposed by David Marr as the *Principle of Least Commitment* [81].

A Machine Learning Method As mentioned previously,[1] many organizations have a large inventory of printed CAD drawings for complex systems for which no electronic CAD exists. A major goal is the automatic retrieval of important information from the drawing so as to populate a database which can then be

[1]This material is taken from a project done jointly by Thomas C. Henderson, Anshul Joshi, Srivishnu Satyavolu, and Wenyi Wang.

used to query the design (for example, in a reverse engineering problem on legacy equipment). A method is described to retrieve the information found in the legend of an engineering drawing raster image. This proceeds as follows:

- *Segment Legend*: this consists of the segmentation of the set of boxes (the legend form) that contain the information about the drawing; e.g., part name, part number, designer, date, etc.
- *Segment Text*: the text strings must be segmented from within each box of the legend.
- *Segment Characters*: individual characters must be segmented from the text strings.
- *Classify Characters*: each character must be labeled as to its semantic class.

The method depends on a semantic approach to box extraction, followed by a variation on standard string segmentation, and character classification is performed using Radial Basis Functions, augmented by a logical decision method between similar letters which are confused. The major goal is to take large sets of engineering drawing images and populate a database with the information from the legends of the drawings as automatically as possible. We describe the algorithms and the experiments performed using this method.

We break the process into several sub-tasks. Generally, there are a variety of possible engineering drawing types, including general arrangement drawings, arrangement drawings, assembly drawings, detail drawings, and fabrication drawings. The first task is to determine which types of drawings exist in the dataset and their formats. A typical example is the detail drawing shown in Fig. 3.2. The solution requires the identification of the *form* (i.e., the specific layout of lines and text structure of the legend), and this, in turn, is based on the extraction of text and straight lines in specific areas of the drawing. For example, in Fig. 3.2, the name of the drawing (the part name, in this case) is "RETAINER, PACKING" and the Part No. is 8737863. To find the legend with the name and part number requires the extraction of text and straight lines with specific relations holding between them. The drawing has a regular structure which can be exploited to find the legend (a set of boxes in the lower right corner), as well as to interpret it.

Figure 3.10 shows the engineering drawing analysis process used here. First, the boxes are recovered from the image. This is done by analyzing the vertical and horizontal line segments. Once an initial set is extracted, then gaps are filled where appropriate, and rectangular box structures are produced. If a set of sufficient number (i.e., > 30) of boxes is found in the lower right corner of the image, then a neighborhood connectivity graph is formed. A best match correspondence between this graph and the template legend graph is determined. From this, the appropriate boxes are selected in order to find the part name, part number and other information of interest. This is done through the use of Radial Basis Functions which are developed based on a set of characters selected from the engineering drawing images.

A novel box segmentation algorithm has been developed, and its operations are shown in Fig. 3.11. In the first step, the maximum horizontal and vertical traversable

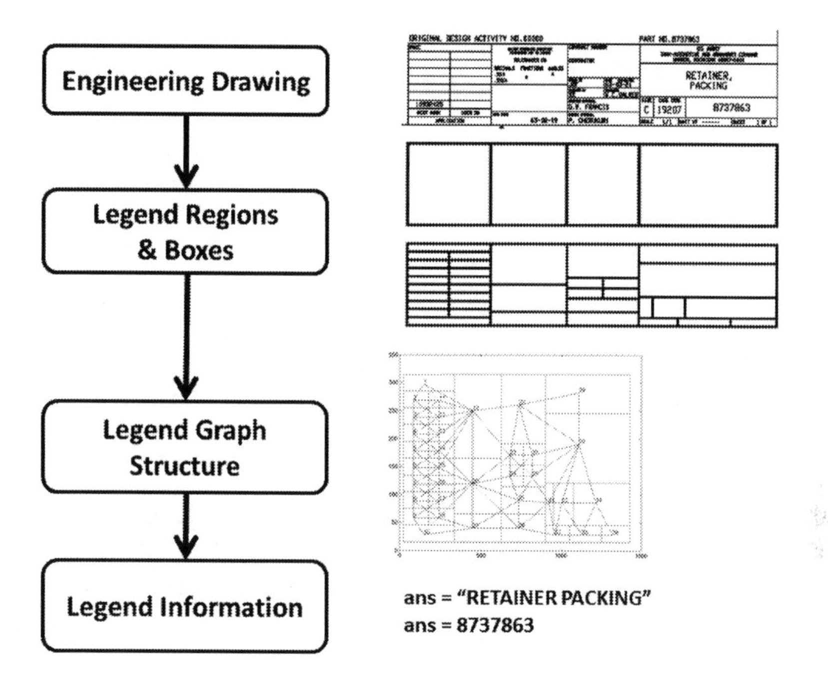

Fig. 3.10 The engineering drawing analysis sequence

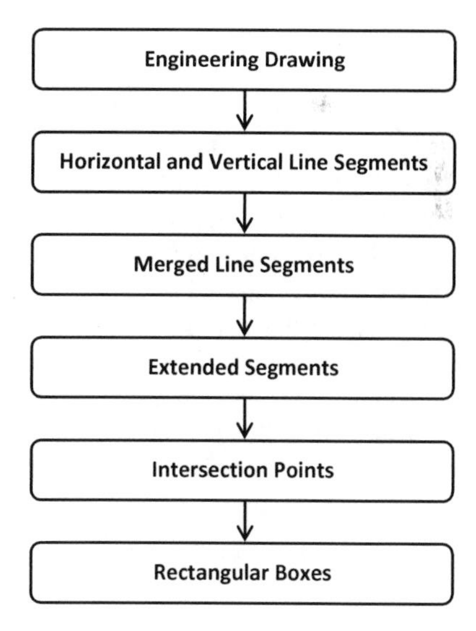

Fig. 3.11 The box analysis sequence

distance from each foreground pixel is found and used to select interest pixels. The connected components of these are labeled, and appropriate neighbors are merged; this means that their ends are nearby and that there is no perpendicular cutting segment between them. Next, these segments are extended out from the

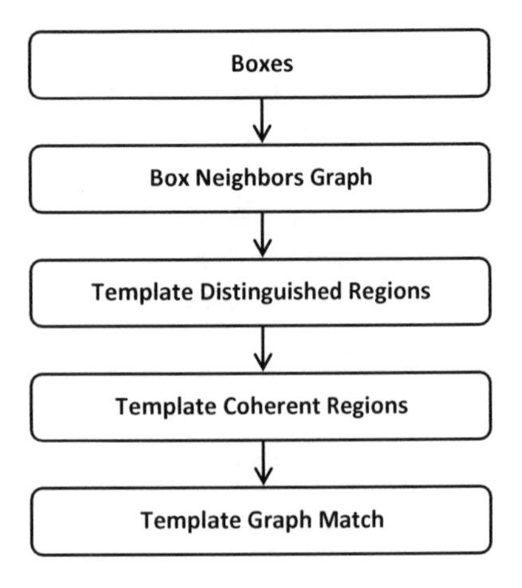

Fig. 3.12 The graph template analysis sequence

endpoints as far as foreground pixels exist in the original image. To get the boxes, the intersection points of the horizontal and vertical segments are located; then each point is considered as an upper left corner point and appropriate other corners sought. Occasionally there is a gap in a line segment which prevents the formation of a box. We perform a post box scan for segments which mostly traverse a box and make a new box for those.

The neighborhood relations are key to understanding the semantics of the boxes as they define the legend; therefore, a graph structure is imposed on the boxes. For each box, the neighbors (i.e., if there exist any neighboring pixels), as well as their spatial layout (up, left, right or down) is determined. Figure 3.12 shows the steps in the graph template matching algorithm. The distinguished template regions are vertical regions with the same set of boxes comprising each column, whereas the coherent template regions are those with a vertical edge through the height of the box region. Figure 3.13 shows the results of this analysis applied to a test image; the top image shows the boxes, the middle shows the distinguished template regions, and the bottom image shows the final coherent template regions found. These latter correspond to a semantic grouping of legend information. Figure 3.14 shows the graph structure recovered for this test image. Note that Box 30 has the part name and Box 34 has the part number. These boxes are matched due to the uniqueness of their spatial neighborhood layout. E.g., Box 30 has the following graph information (taken from Matlab):

```
>> im53r_info.nodes(30)
ans =
        nei: [22 27 28 29 31 33 34]
      right: []
         up: [22 29]
       left: [22 27 28]
       down: [28 31 33 34]

>>
```

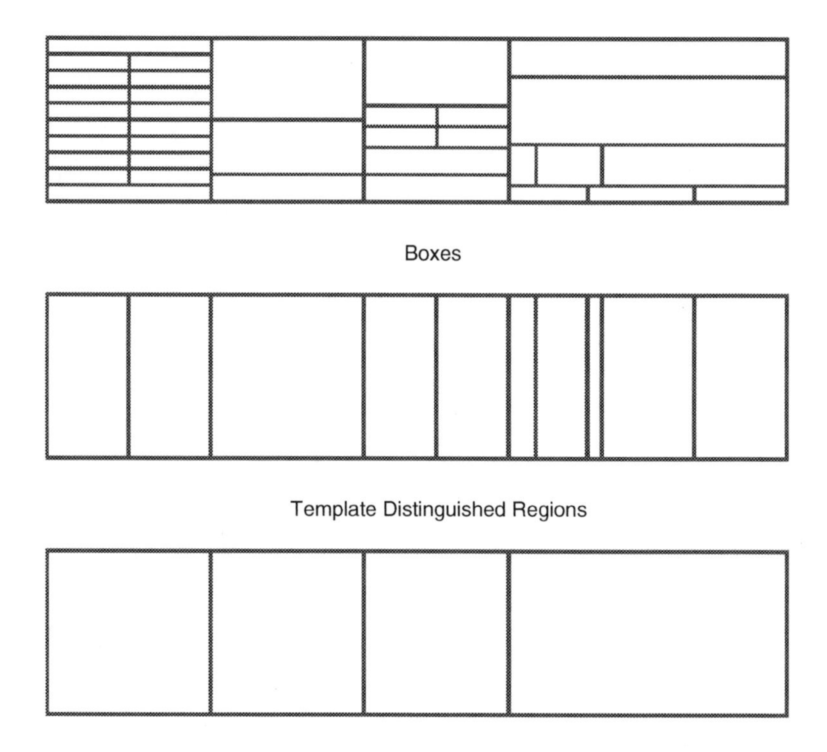

Fig. 3.13 Results of the graph template analysis of test image

As can be seen, Boxes 22 and 29 are above (i.e., have pixels above the minimum row of Box 30), Boxes 22, 27, and 28 are to the left and Boxes 28, 31, 33 and 34 are down from it. At this point we merely use 1-hop neighbors, but for graphs with similar nodes, multiple hops could be used to distinguish nodes.

The final step in the information retrieval from engineering drawing legends is to segment the text and classify the characters. The specific goal here is to extract the part numbers and part names. We use radial basis functions (Marsland's formulation [82]) to classify characters. These are a form of neural network which uses training data as local neurons in the input space, and computes a distance function from the input, followed by a perceptron layer which selects the character. A single class identifier is used for each digit and uppercase letter (i.e., there are 36 classifiers and the max response is selected as the class). Each character image is resized to a standard 30x20 image; next the skeleton of the character is found, followed by a dilation by 1. The resulting image is converted to a 1x600 vector which represents the component. The RBF is first trained and then applied. The training was done with 9 sample images of each character. In a leave-one-out train and test, we had 897 out of 900 (99.67%) correct for digits, and 6058 of 6084 (99.57%) for upper case letters.

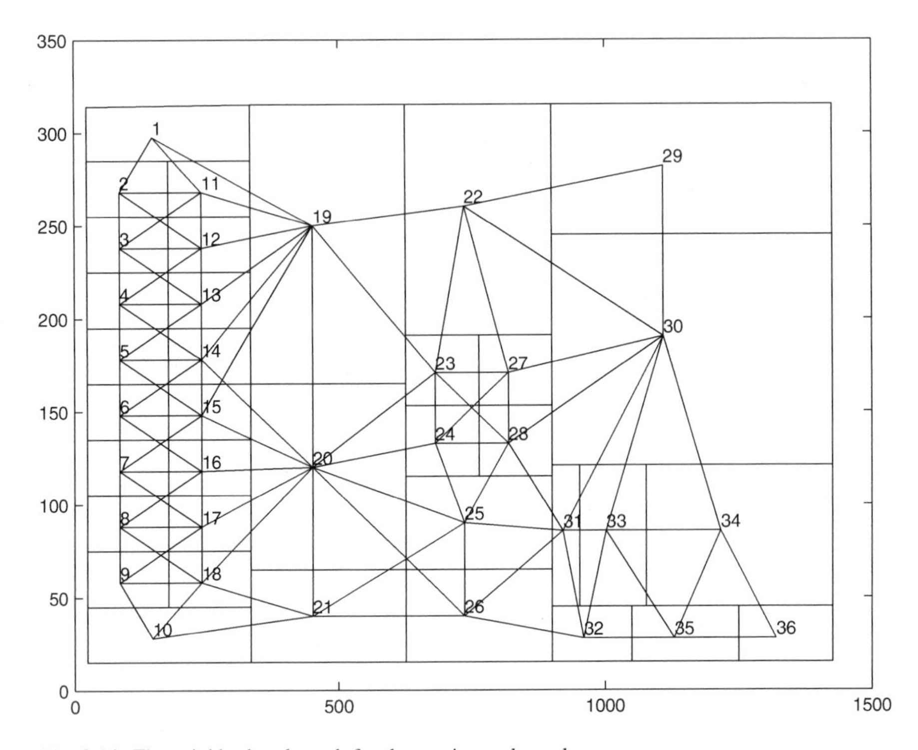

Fig. 3.14 The neighborhood graph for the test image legend

Training

- Set RBF centers to selected sample points
- Calculate Gaussian distance

$$(g(x, w, \sigma) = exp\left(\frac{-\parallel x - w \parallel^2}{2\sigma^2}\right)$$

- Train output weights by computing pseudo-inverse of the activations

Application simply involves passing an unknown vector through the network The quality of the scanned images in our particular application guarantees that for the most part, individual characters correspond to single connected components. Thus, the overall algorithm takes the given box, finds the connected components in the box, sorts them as to top-down line and left to right text structure.

This technique was developed using 16 engineering drawing images, like the example shown in Fig. 3.2. Of the 262 letters in the part names, all were correctly classified, except for a broken letter "N" that was classified as a "W"—thus achieving over 99% accuracy. Figure 3.15 shows two bad characters: the mis-classified "N" and a correctly classified broken "P". We then tested the method on

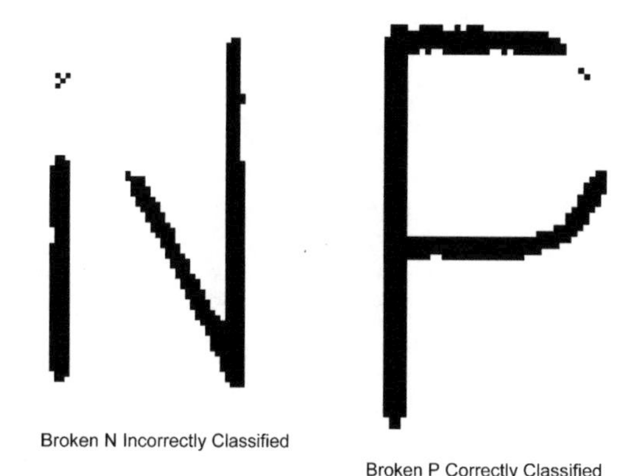

Fig. 3.15 Some bad characters and the results

10 new images, and of those, one had been scanned with a wrinkle and failed (the horizontal lines were not straight), but the classification accuracy on the remaining nine images was 97%. The misclassified letters included "D" for "O" (2), "D" for "S" (1), "L" for "E" (1), and "D" for "B"; (1). When applied to extracting the part numbers, the algorithm achieved 100% accuracy on the training set, and 100 % accuracy on the part numbers.

In this approach then: (1) a technique is proposed and demonstrated for matching box structure graphs to a template, (2) the box structure is exploited to constrain the text extraction from important boxes (e.g., part name, part number, etc.), (3) a Radial Basis Function classifier (along with confounded letter recognition logic) is developed which achieves 100% accuracy for digits in the part number and over 97% accuracy for letters in the part name.

Current areas of research consider the use of machine learning techniques to automatically extract the box structure in a set of engineering drawings; in addition, it may be possible to use standard drawing conventions where they have been used. The application of the RBF method to lower case letter is also an interesting problem; this is extremely difficult in many cases because the lower case letters are run together and of low resolution. Finally, it may be possible to use constraint-based techniques to achieve a more efficient subgraph isomorphism method; e.g., this could be based on previous work in this area (see [43, 87]).

Chapter 4
A Structural Model for Engineering Drawings

We[1] have proposed a structural modeling approach combined with a nondeterministic agent system to produce interpretations of scanned images of CAD drawings [47–50, 113]. This allows a broad set of thresholds to be used during the image analysis which in turn permits the efficient pruning of the resulting search space by taking advantage of constraints from the model or specific application. The nondeterministic agent system is described in a Chap. 5.

In order to organize the activity of the analysis agents, we have developed an engineering drawing model comprised of structures typically found in such drawings, and relations between those structures. This approach is based on structural and syntactic shape methods (e.g., see [52]); however, our method is novel in that it allows for the natural application of the NDAS agent system to recover the desired structures from an image.

4.1 Terminal Structures

Terminal structures correspond to the terminal symbols of a shape grammar, and include: **text**, **box**, **pointer_ray**, **pointer_line**, **pointer_arc**, **circle**, **line_segment**, and **graphic**. Technically, these are defined as follows (the terminals and nonterminals have been numbered – given in parentheses):

- (1) **text**: characters in the image (aligned and spatially close).
- (2) **box**: closed rectangular set of 4 line segments at 90 degrees in sequence.
- (3) **pointer_ray**: line segment with arrow at one end (may be polyline of degree 2).
- (4) **pointer_line**: line segment with arrow at each end.
- (5) **pointerarc_ray**: arc segment with arrow at one end.

[1] This chapter is contributed by Lavanya Swaminathan; it is Chap. 3 of her MS thesis.

T.C. Henderson, *Analysis of Engineering Drawings and Raster Map Images*,
DOI 10.1007/978-1-4419-8167-7_4,

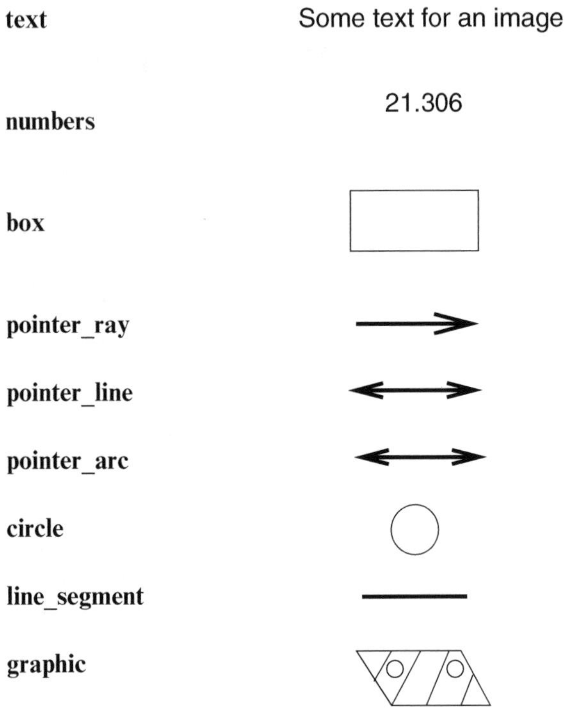

Fig. 4.1 Examples of terminal structures

text Some text for an image

numbers 21.306

box

pointer_ray

pointer_line

pointer_arc

circle

line_segment

graphic

- (6) **pointerarc_line**: arc segment with arrow at each end.
- (7) **circle**: image segment which is a circle.
- (8) **line_segment**: image segment which is a straight solid line.
- (9) **graphic**: set of image segments grouped together and classified as graphic elements.
- (10 – 20) unused.

Terminal structures are discovered by specific analysis agents which examine both the image and outputs from other agents. Figure 4.1 shows a labeled instance of each terminal structure.

4.2 Higher-Level Structures

Higher-level structures correspond to the nonterminal symbols in a shape grammar, and can be described by rewrite rules which define sub-structures which comprise the new structure, and the relations that must exist between the sub-structures. Loosely described, these are given as:

- (21) *pointer_ray1*: enforces distinct instance of **pointer_ray**.
- (22) *pointer_ray2*: enforces distinct instance of **pointer_ray**.

- (23) *pointerarc_ray1*: enforces distinct instance of **pointerarc_ray**.
- (24) *pointerarc_ray2*: enforces distinct instance of **pointerarc_ray**.
- (25) *line_segment1*: enforces distinct instance of **line_segment**.
- (26) *line_segment2*: enforces distinct instance of **line_segment**.
- (27) *line_segment3*: enforces distinct instance of **line_segment**.
- (28) *text1*: enforces distinct instance of **text**.
- (29) *text2*: enforces distinct instance of **text**.
- (30) *symmetric_pointer_pair_in*: 2 collinear **pointer_rays** with arrows at near ends. These will point inward toward dimension lines.
- (31) *symmetric_pointer_pair_out*: 2 collinear **pointer_rays** with arrows at distant ends. These will point outward toward dimension lines.
- (32) *symmetric_pointerarc_pair_in*: 2 **pointerarc_rays** with arrows at near ends. These will point inward toward dimension lines.
- (33) *symmetric_pointerarc_pair_out*: 2 **pointerarc_rays** with arrows at distant ends. These will point outward toward dimension lines.
- (34) *dimension_rays_in*: represents **text** and inward pointing rays of the dimension construct.
- (35) *dimension_rays_out*: represents **text** and outward pointing rays of the dimension construct.
- (36) *dimension*: **text** with a *symmetric_pointer_pair*.
- (37) *dimension_angle_rays_in*: **text** and inward pointer rays of angle description.
- (38) *dimension_angle_rays_out*: **text** and outward pointer rays of angle description.
- (39) *dimension_angle*: either of (37) or (38).
- (40) *dimension_set*: a *dimension* enclosed by a *symmetric_line_pair* or an *angle_dimension* enclosed by an asymmetric_line_pair.
- (41) *angle_set*: pair of connected **line segments** with angle_arrow between them.
- (42) *pointer_ray_extn*: a **pointer_ray** extended by a **line_segment**.
- (43) *pointer_line_extn*: a **pointer_line** extended by a **line_segment**.
- (44) *pointerarc_line_extn*: a **pointerarc_line** extended by a **line_segment**.
- (45) *pointer_line_extn_in_circle*: an extended **pointer_line** embedded in a **circle**.
- (46) *check_sign*: ($\surd$) : 2 **line segments** arranged as a check mark. These occur in drawings to label various aspects.
- (47) *check_pair*: *check sign* with **text**.
- (48) *dimension_description*: a *dimension_set* and corresponding **graphic**. Complete dimension information, including the graphic being described.
- (49) *text_in_box*: **text** in a **box**.
- (50) *text_in_box1*: enforces distinct instance of *text_in_box*.
- (51) *text_in_box2*: enforces distinct instance of *text_in_box*.
- (52) *text_in_box3*: enforces distinct instance of *text_in_box*.
- (53) *text_in_box4*: enforces distinct instance of *text_in_box*.
- (54) *one_datum_ref*: 3 collinear and adjacent *text_in_box*.
- (55) *two_datum_ref*: 4 collinear and adjacent *text_in_box*.
- (56) *datum_ref*: either of (54) or (55).
- (57) *datum_below_text*: a **datum_ref** below **text**.

- (58) *dashed_lines*: 3 collinear **line_segments** specifying a cross-section in the image.
- (59) *dash_line1*: enforces distinct instance of *dashed_line*.
- (60) *dash_line2*: enforces distinct instance of *dashed_line*.
- (61) *circle_center_dim*: 2 **dashed_lines** perpendicular to one another intersecting on the center of the **circle** they describe.
- (62) *only_graphics*: only **graphics** after eliminating all components of dimension.
- (63) *text_comb*: combination of 2 **texts**.
- (64) *text_final*: enforces distinct instances of **text** or **text_comb**.
- (65 – 80) unused.

4.2.1 Relations

In order to define these structures, a number of (mostly) geometrical relations must be defined between sub-structures and determined to exist in an image:

- *inBox(x,***box***)*: x is contained within a **box**.
- *inCircle(x,***circle***)*: x is contained within a **circle**.
- *touches(x,y)*: x and y are part of same image segment.
- *adjacent(x,y)*: x and y are adjacent to each other.
- *parallel(x,y)*: x and y have parallel axes.
- *perpendicular(x,y)*: x and y have perpendicular axes.
- *near(x,y)*: x and y within some distance.
- *collinear(x,y)*: major axes of x and y on same line.
- *length(x)*: length of x.
- *angleBetween(x,y)*: angle between x and y.
- *between(x,y,z)*: y is between x and z.
- *horizontal(x)*: x is horizontal.
- *unequal(x,y)*: x and y are not the same.
- *bisect(x,y)*: x and y are **line_segments** that bisect one another.
- *above(x,y)*: y is above x.
- *below(x,y)*: y is below x.
- *touchesEnd(x,y)*: an end of x touches an end of y.

4.2.2 Productions (Rewrite Rules)

We now give a more technical definition of the structures and relations.

- **terminal structures**: these are defined in Sect. 4.1, and are found by the respective agents when applied to the various images and outputs available from other image analysis agents.

- **relations**: geometrical relations between structures.

$$inBox(A, box) \Rightarrow centroid(x) in\ box$$

$$inBox(A, box) \equiv min(x, y)box \leq min(x, y)A$$

$$\wedge max(x, y)A \leq min(x, y)box$$

$$inCircle(A, circle) \Rightarrow centroid(x) in\ circle$$

$$inCircle(A, circle) \equiv min(x, y)circle \leq min(x, y)A$$

$$\wedge max(x, y)A \leq min(x, y)circle$$

$$touches(x, y) \equiv segment(x) = segment(y)$$

$$parallel(x, y) \equiv \angle x, y = 0 \pm \varepsilon$$

$$perpendicular(x, y) \equiv \angle x, y = \frac{\pi}{2} \pm \varepsilon$$

$$near(x, y) \equiv distance(x, y) \leq \varepsilon$$

$$collinear(x, y) \equiv distance(line(x), line(y)) \leq \varepsilon$$

$$same\_length(x, y) \equiv length(x) = length(y) \pm \varepsilon$$

$$between(x, y, z) \equiv l = line(x, z) \wedge proj(b, l) between\ x, z$$

The rewrite rules describe how higher level structures are derived from lower level ones. See Appendix A for the complete set of rewrite rules. Table 4.1 gives these rewrite rules as a simple table by symbol number.

4.3 Goal Graphs

Just as for image analysis, the agents which determine structure in the drawing are activated by the existence of certain triggering data. This is organized by way of goal graphs, which are generated by using the grammatical description to formulate a constrained search process. This works bottom-up by resolving the terminal structures as parts of higher-level structures.

First, some definitions:

A **ground structure** is a structure arising from the analysis of the image.
An **instance structure** is a ground structure or the result of a satisfied goal graph analysis.
A **model structure** is a structure arising from a rewrite rule.
A **goal graph generator** is an instance structure which gives rise to a goal structure.
A **goal graph** consists of the model structures and relations between them defined by the right hand side of a rewrite rule.
A **bound instance structure** is an instance structure assigned to a model structure in a goal graph.

Table 4.1 Rewrite rules

Rule	LHS	RHS values	Rule	LHS	RHS values
1	21	3	2	22	3
3	23	5	4	24	5
5	25	8	6	26	8
7	27	8	8	28	1
9	29	1	10	64	1
11	63	28 29	12	64	63
13	30	21 22	14	31	21 22
15	32	23 24	16	33	23 24
17	34	30 64	18	35	31 64
19	36	35	20	36	34
21	37	32 64	22	38	33 64
23	39	38	24	39	37
25	40	25 26 36	26	41	25 26 39
27	42	8 3	28	43	8 4
29	44	8 6	30	40	25 26 44 64
31	45	43 7	32	46	25 26 27
33	47	46 64	34	48	64 42 8
35	49	1 2	36	50	49
37	51	49	38	52	49
39	53	50 51 52	40	54	53
41	55	64 54	42	48	8 42 55
43	48	64 45 8	44	48	45 55
45	56	25 26 27	46	57	56
47	58	56	48	59	57 58
49	60	9 40	50	48	40 60

A goal graph is **satisfied** if every model structure has a bound instance structure and all relations are satisfied. Goal graphs can be formed in a bottom-up or a top-down fashion as described below. Here we use only the bottom-up approach.

4.3.1 Bottom-up Formation of Goal Graphs

Algorithm Goal Graph Up:
On input: instance structures
On output: goal graphs
A1. Every instance structure gives rise to a goal graph for every rewrite rule in which it appears in the right hand side.

A2. The goal graph generator is associated with its model structure counterpart in the goal graph.

A3. A right hand side analysis agent attempts to find assignments of other instance structures to the remaining goal graph structures that satisfy all the relations of the right hand side.

A4. If instance structures are found which can be assigned to model structures in the goal graph and which satisfy the relations, then the left hand side model structure is added as an instance structure, and the resulting satisfied goal graph is output.

4.3.2 Top-down Goal Graph Analysis

Once the bottom-up application of these rules has terminated, if there are any ground structures which are not part of a description structure, then the following rules may be applied.

Algorithm Goal Graph Top-Down:
On input: goal graphs not satisfied; ground structures
On output: satisfied goal graphs
B1. Every unsatisfied goal graph is assigned to a top-down search agent.

B2. A search is made to discover in the remaining instances any unbound ground structure satisfying the right hand side relations with respect to the bound instance structures.

A **goal graph process** is the application of Algorithm Goal Graph Up to a set of ground structures from an image.

The **coherence** of a goal graph process is the average number of satisfied goal graphs generated by each instance structure.

There are several ways to enforce coherence. In particular, a limit (called the generation limit) can be placed on how many goal graphs an instance structure can generate. This can be context independent or not. For example, the generator limit may be a function of depth in the rewrite rules. This may also be controlled by varying the thresholds on the relations; e.g., tighter constraints will yield fewer interpretations.

4.4 Analysis Complexity Reduction

Given a strategy of generating as much of the search space as possible, it is necessary to find ways to reduce the number of alternatives; however, this must be done in a systematic and correct way. We have developed two strategies:

- *Symbolic Pruning*: exploit the formal aspects of the grammar and the algebraic relations to eliminate duplications and redundancies.
- *Empirical Pruning*: exploit the specific statistics and findings resulting from samples taken from a given domain.

The next sections develop the framework for these two approaches.

4.4.1 Symbolic Pruning

Our goal is to determine the complexity of the total number of possible symbols generated from a set of ground structures with respect to a specific grammar. Let G be a grammar, and for the current context, let it represent the rewrite rules. We define a *production sequence* as a correct application in some order of the rewrite rules of G. Thus:

$$ps = \prod_{j=1}^{n} i_j$$

defines a production sequence, ps, where i_j is the index of the jth production (repeating an index is allowed).

Symbol redundancy of a vocabulary symbol, v, called $SR(v)$, is the count of the number of distinct production sequences that produce the symbol. This is the same as the number of ways the ground structures can be mapped onto the terminal symbols to produce the symbol v. For a terminal symbol, a, we have

$$SR(a) = |g|$$

where $|g|$ is the number of ground structures of this terminal symbol.

In order to calculate $SR(v)$ for a non-terminal symbol v, we introduce the following notion. An **0-form rewrite rule** is one with only terminal symbols on the right hand side. An **0-form grammar** is one with only 0-form rewrite rules.

Algorithm 0-form produces an 0-form grammar from a general grammar.

```
Algorithm 0-form

On input: a general grammar, G
On output: an 0-form grammar, G2
G1 <- G
G2 <- empty set

whenever there exists a rewrite rule, R, in G1 such
that lhs(R) is a non-terminal and there are only
terminals in the rhs(R), then:
     Add R to G2
```

```
For every R' in G1 such that lhs(R) is in the
rhs(R')
Replace each occurrence of lhs(R) in R' with
rhs(R)
and call new rule R''
Add R'' to G1
```

Note that this applies to non-recursive grammars, but if there is a recursive rewrite rule, then it can be easily flagged, and the user can set a maximum depth for it. For example, any number of section lines can occur in a technical drawing so long as they are parallel.

For a terminal symbol w and a rewrite rule R, define **count(w,R)** to be the number of times that w appears on the right hand side of R. Then, for a rewrite rule in 0-form and a non-terminal symbol, v, we have:

$$SR(v) = \sum_{R} \prod_{w} [SR(w)(SR(w) - 1) \ldots (SR(w) - count(w, R) + 1)]$$

where R is in the rewrite rules, and the sum is taken over all rules with v in the left hand side, and w is a distinct symbol in the right hand side of R. If any summand is negative or zero, then it is not added in; if all summands are negative or zero, then v cannot be produced.

Consider the following example grammar (disregarding specific relations between symbols – assume that all are symmetric):

```
G = { PPS -> SLP1 + SLP2
      SLP1 -> SLP
      SLP2 -> SLP
      SLP -> line1 + line2
      line1 -> line_segment
      line2 -> line_segment
}
```

The $G1$ set produced by Algorithm 0-form is:

```
G1 = { PPS -> SLP1 + SLP2
       SLP1 -> SLP
       SLP2 -> SLP
       SLP -> line1 + line2
       line1 -> line_segment
       line2 -> line_segment
       SLP -> line_segment + line2
       SLP -> line_segment + line_segment
       SLP1 -> line_segment + line_segment
       SLP2 -> line_segment + line_segment
       PPS -> line_segment + line_segment + SLP2
       PPS -> SLP1 + line_segment + line_segment
       PPS -> line_segment + line_segment
```

```
            + line_segment + line_segment
}
```

The *G*2 set produced by Algorithm 0-form is:

```
G2 = { line1 -> line_segment
       line2 -> line_segment
       SLP -> line_segment + line_segment
       SLP1 -> line_segment + line_segment
       SLP2 -> line_segment + line_segment
       PPS -> line_segment + line_segment
           + line_segment + line_segment
}
```

Consequently, we have the following results:

```
Suppose that    SR(line_segment)   = 4

then            SR(line1)  = 4
                SR(line2)  = 4
                SR(SLP)    = 12
                SR(SLP1)   = 12
                SR(SLP2)   = 12
                SR(PPS)    = 24
```

Note that the symbolic redundancy depends on the number of ground structures. For example:

```
Suppose that    SR(line_segment)   = 2

then            SR(line1)  = 2
                SR(line2)  = 2
                SR(SLP)    = 2
                SR(SLP1)   = 2
                SR(SLP2)   = 2
                SR(PPS)    = 0    fails!
```

There are not enough ground structures to satisfy the need for four distinct terminal symbols to produce nonterminal symbol *PPS*.

The symbol redundancy measure is a worst case estimate since for a specific set of ground structures, the required relations between the terminal symbols may not hold; this would prevent the generation of the nonterminal symbol. However, this does give a useful measure of the complexity of the analysis to be performed (and, in fact, makes it possible to know when there are too few ground structures to produce a complete parse).

4.4.2 Exploitation of Symbol Redundancy

The symbol redundancy measure of the number of ways in which a given symbol can be produced allows an analysis to determine if any of the redundant production sequences can be eliminated. There are two approaches to actually eliminate redundancies:

- compile out one rewrite rule: start symbol, with relations between terminals; eliminate combinations where only relations are symmetric; synthesize the relations for this rewrite rule.
- allow only one production sequence from a symmetric relation equivalent set when they differ only in the assignment of the same symbol in the right hand side.

We describe a solution using the second approach. To do this requires the following definition: two production sequences are *symmetric relation equivalent* if they differ only in assignment of same symbol in right hand side of rewrite rule. Likewise, a set of production sequences are symmetric relation equivalent if they are all pairwise symmetric relation equivalent.

One approach to symbolic pruning then is to find sets of symmetric relation equivalent production sequences and to allow only one during the structural analysis. This is a parse time operation.

4.5 Empirical Pruning

While symbolic pruning results from the structural relations that exist in the grammatical description of the shape, it is also possible to make use of parameters (thresholds, angles, etc.) whose values depend on the specific technical drawings being analyzed. We have implemented an interface that allows the user to look at intermediate results from various agents and to assert which threshold or value is the best. Once the user selects a value, or a range of values, this is noted, and the other alternatives are not considered further (i.e., those results are discarded and no agent will use them as inputs).

4.6 A Simple Example: Dimension Sets

Figure 4.2 shows a simple engineering drawing with four dimension sets.

A structural rewrite rule for this dimension set is:

```
dimension_set := ptr_ray1 + ptr_ray2 + text

where
```

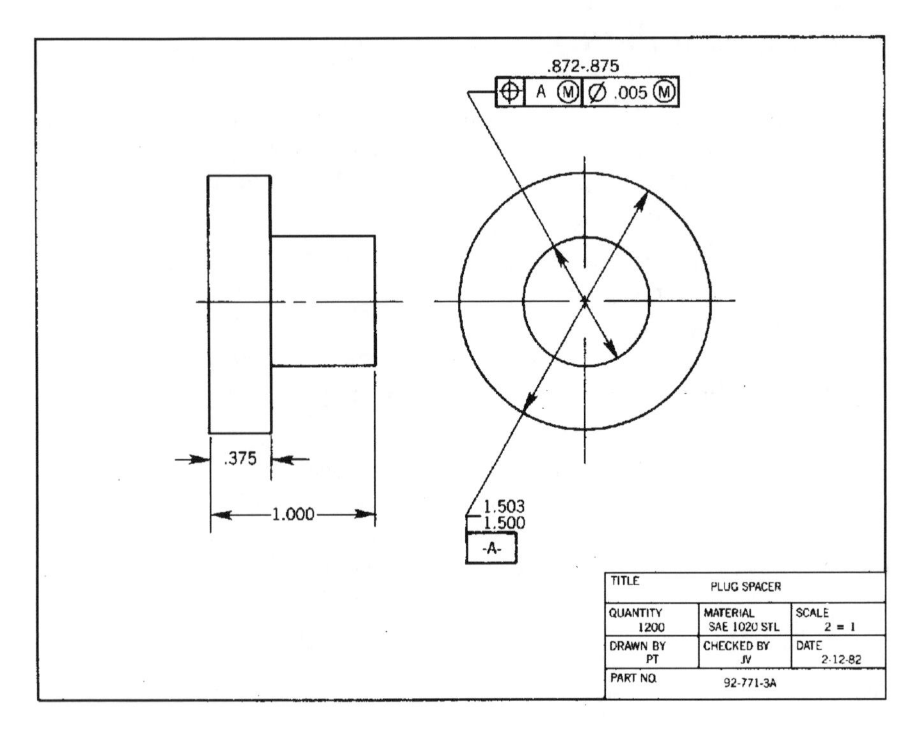

Fig. 4.2 Simple engineering drawing

```
collinear(ptr_ray1,ptr_ray2)
collinear(ptr_ray1,text)
collinear(ptr_ray2,text)
[between(ptr_ray1,text,ptr_ray2)]
```

The analysis described here was performed by a set of nondeterministic agents, and in order to organize their activity, we have developed the engineering drawing model comprised of structures typically found in such drawings, and relations between those structures (this model is described above). This model is novel in that it allows for the natural application of the NDAS agent system to recover the desired structures from images.

The nonterminals used here include:

- *dimension*: **text** with a symmetric pointer pair
- *dimension_set*: a *dimension* enclosed by a *symmetric_line_pair* or an *angle_dimension* enclosed by an *asymmetric_line_pair*.
- *dimension_description*: a *dimension_set* and corresponding **graphic**

We have applied this to the image shown in Fig. 4.2; all four dimension sets were found. Table 4.2 gives the symbol redundancy for the vocabulary symbols

Table 4.2 Rewrite rules

Symbol	Worst case	After pruning
line_segment	21	21
pointer_ray	4	4
text	2	2
circle	1	1
box	1	1
pointer_ray1	4	4
pointer_ray2	4	4
line_segment1	21	21
line_segment2	21	21
line_segment3	21	21
text1	2	2
text2	2	2
text_comb	2	0
text_final	4	2
symmetric_pointer_pair_in	12	0
symmetric_pointer_pair_out	12	2
dimension_rays_in	48	0
dimension_rays_out	48	2
dimension	96	2
dimension_set	40320	2
pointer_ray_extn	84	0
check_sign	420	0
check_pair	1680	0
dimension_description	>100000	2
text_in_box	4	0
text_in_box1	4	0
text_in_box2	4	0
text_in_box3	4	0
text_in_box4	4	0
one_datum_ref	0	0
datum_ref	0	0
datum_below_text	0	0
dashed_lines	7980	3
dashed_lines1	7980	3
dashed_lines2	7980	3
circle_center_dim	>100000	1

in our grammar for engineering drawings. Several examples were run with good success and are reported in [113]. The results of these experiments are encouraging. The structural analysis proceeds correctly, and the system explores much of the interesting part of the search space. The percentage of dimension sets found is 16.6 % for noisy images to 100 % for clean images, and the pruning methods lead to orders of magnitude reductions in the number of symbols considered during the analysis.

Chapter 5
Non-deterministic Analysis Systems (NDAS)

As described[1] in Chap. 1, traditional approaches to drawing analysis follow the process sequence of digitization, noise removal, segmentation, vectorization, recognition of text and graphics, extraction of dimension sets, and semantic analysis using some form of rules. Typically, fixed thresholds are used for each step in the process. As an example of how this process would work ideally, consider the drawing shown in Fig. 4.2. The results of text extraction are shown in Fig. 5.1; Figure 5.2 shows pointer rays from dimension sets; Fig. 5.3 shows pointer lines; Fig. 5.4 shows the circles found in the drawing; Fig. 5.5 shows boxes; and Figs. 5.6–5.9 show the dimension sets in the drawing.

5.1 An Agent Architecture

Agents are independent software processes with the following properties:

1. autonomous (react to environment)
2. have state (beliefs, commitments, etc.)
3. persistent (process never terminates)
4. can communicate (send and receive messages related to effort)
5. perform some action (have abilities to analyze and create data).

For more complete accounts, see [112, 129].

An agent architecture is a software architecture for decision making with intelligent (flexible) processes embedded within it. The agents may be proactive or reactive, and should cooperate (including communicate) to achieve a goal.

We explore the use of nondeterministic agent systems (NDAS) to achieve a more flexible system for technical drawing analysis. (see Appendix B for a set of image analysis agents). They are called nondeterministic because the agents explore

---

[1] This chapter is contributed by Lavanya Swaminathan based on her MS thesis.

T.C. Henderson, *Analysis of Engineering Drawings and Raster Map Images*,
DOI 10.1007/978-1-4419-8167-7_5,

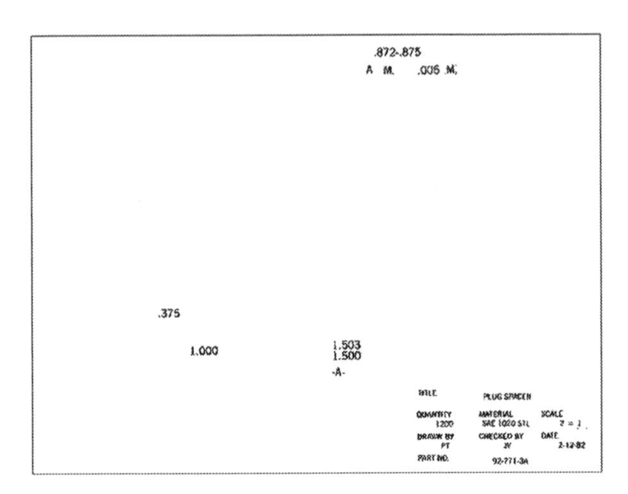

Fig. 5.1 Text

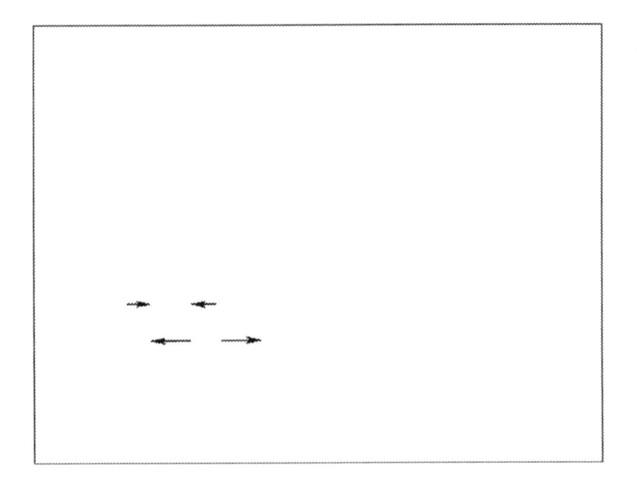

Fig. 5.2 Pointer rays

alternative parts of the solution space simultaneously, and every agent works to produce some result which may or may not contribute to the final result. (Note that this is a form of *speculative parallelism* [17]; this can also be viewed as a distributed blackboard system [130].) The final result derives from only a subset of the work put in by all the agents. We explore nondeterminism in this problem domain since deterministic systems usually make irrevocable decisions (e.g., threshold selection) that eliminate possible solutions. The technical drawing problem domain contains many factors that vary with the drawing: thresholds, text fonts and size, noise levels, etc., and this variation makes it interesting to explore the possible solution space dynamically and in a breadth-first way.

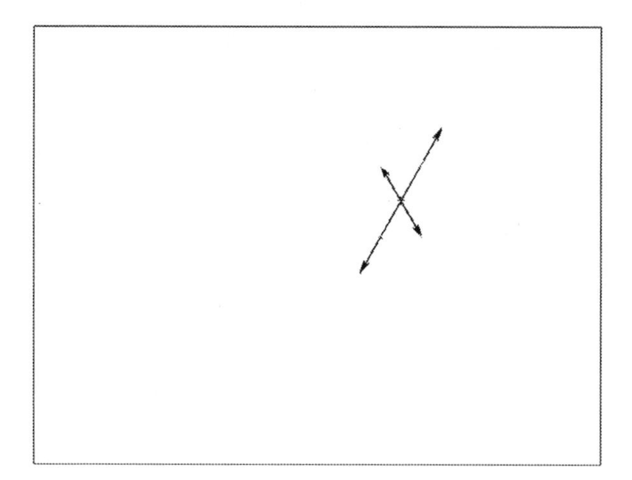

Fig. 5.3 Pointer lines

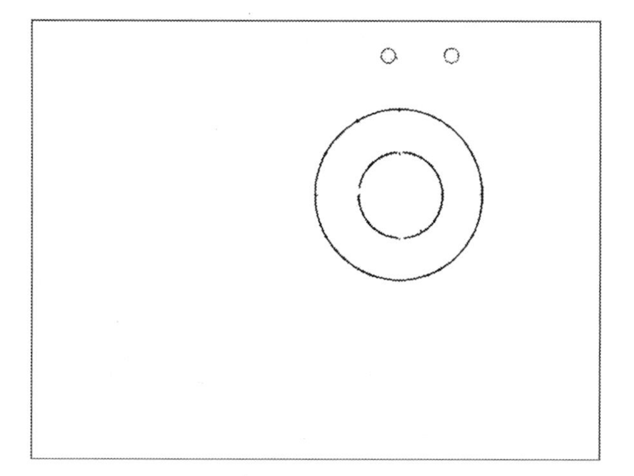

Fig. 5.4 Circles

The issues that we have explored using the NDAS approach include:

- *The analysis of technical drawing annotations*: This includes the ability to recognize dimensioning, features and their annotations, tolerances, and references to nomenclature.
- *threshold sensitivity analysis*: The clear determination of the relationship between a change in threshold and its impact on the analysis process and result.
- *precision, robustness and performance analysis*: An engineering account of these aspects of a system is extremely useful in making cost performance tradeoff decisions. Most results in the literature do not define very carefully what it means for a system to work nor what error is.

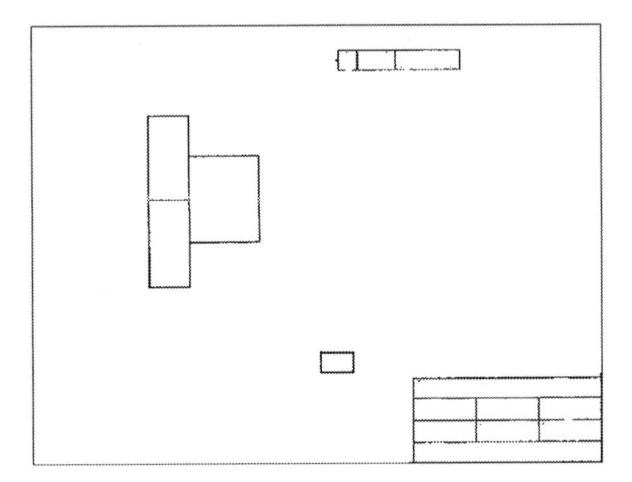

Fig. 5.5 Boxes

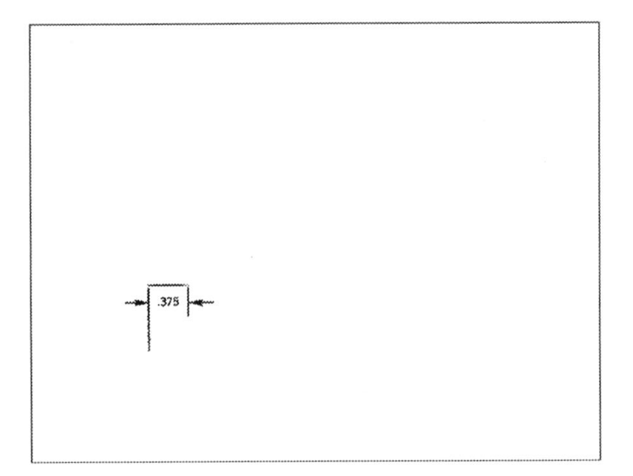

Fig. 5.6 Dimension 1

Annotation models have been presented in the literature (e.g., see [25,28,31,42]), and we have extended these ideas and developed a novel approach to high-level modeling. We have also implemented the basic image analysis tools to extract text, graphics, graphical primitives. These form the basis actions for the agent architecture approach. The thesis of this work is that a structural analysis model may be realized through a set of software agents acting independently and in parallel to ultimately produce a coherent analysis. This has been demonstrated through the design and analysis of the NDAS system and experimental results provided to support the claims.

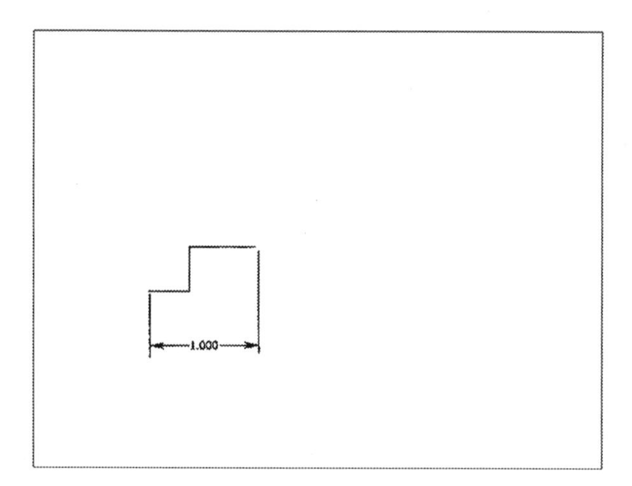

Fig. 5.7 Dimension 2

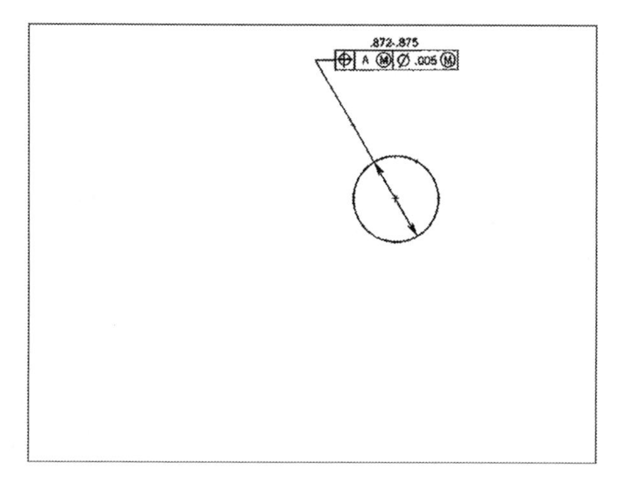

Fig. 5.8 Dimension 3

In terms of NDAS, we have explored the:

1. organization
2. communication, and
3. higher-level modeling capabilities

NDAS system using the analysis of technical drawings as the application domain. Figure 5.10 shows the organizational and operational view of the technical drawing analysis agent system.

The a_i, a_j and a_k's are agents which look for certain kinds of objects in their field of action, and when the appropriate conditions hold, the agent will act and produce another set of objects. These objects may be desired data objects (e.g., a segmented

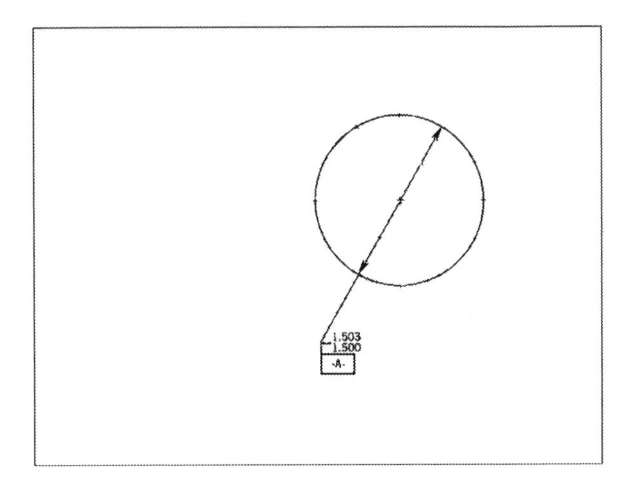

Fig. 5.9 Dimension 4

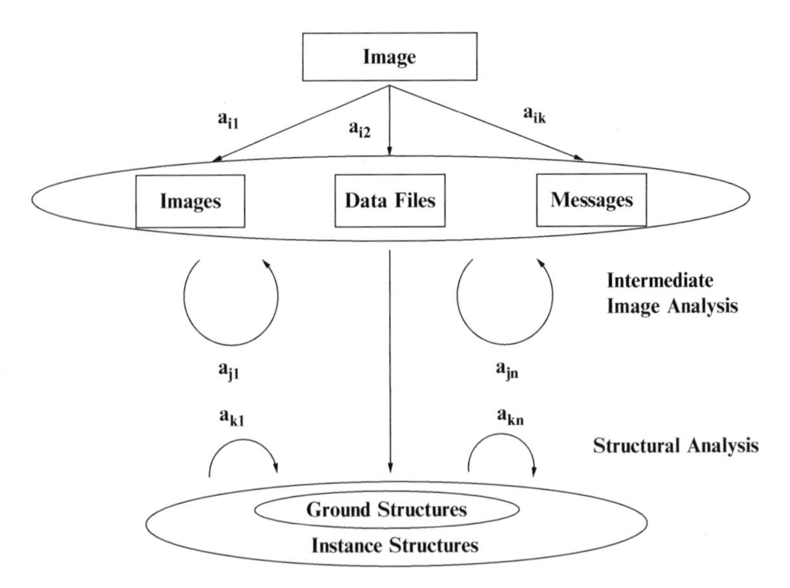

Fig. 5.10 Nondeterministic agent system for technical drawing analysis

image) or a message to enlist new agents (e.g., to produce a better analysis). The idea is that the agents work independently and in parallel and continue to work until conditions cause them to become quiescent.

Such an approach allows for feedback loops as previously processed data may reappear after modification by a higher-level process. This also allows a coarse to fine improvement as more work of broader context is completed. Agents may also monitor the activity of other agents and measure performance or provide data for consumption by the (human) user. Of course, the user may participate as an agent as well! (Note that it is straightforward to define the agents so as to capture the standard processing paradigm.) This approach is based on distributed agents and files, and avoids centralized control or bottlenecks through centralized agents which record and relay information.

A similar structure was proposed by Carriero and Gelernter [17] and called a process trellis. It was a hierarchical graph of decision processes, from low-level to high-level and with others in between. All processes executed continuously and concurrently. This was an example of their specialist—parallel problem approach. The big question was efficiency. The process trellis clearly allows massive parallelism and scales well. Some differences with what we propose are that the trellis:

- has dataflow-like streams of input and output; pass data up and query down,
- there is a master and n worker processes, and
- the goal was a real-time algorithm for medical equipment control.

5.2 NDAS Organization

The nondeterministic agents are a collection of independent Unix processes. There are two aspects of agent organization: internal and external.

5.2.1 Internal Agent Organization

Each agent is written as a C++ program, which watches its current executing environment (directory) for the existence of certain files or processes and then takes actions as specified by its internal code. This can be described by the following set of actions:

1. *Monitor*: watch for the existence of files, processes, etc.
2. *Action Program*: Executable actions to take.
3. *Wrapup*: Cleanup files, check termination conditions, etc.

(For a particular internal agent architecture pseudo-code layout, see Appendix B.)

5.2.2 External Organization

The organization between agents includes how they announce their existence, if necessary, file protocols, etc. In addition, this includes the variety of agents that the user may wish to define; e.g., facilitation agents, performance analysis agents, process control agents, image analysis agents, feature analysis agents, higher-level model agents, etc.

Communication between agents is also an issue of their organization. We have them communicate through files; for example, each agent creates a file which describes itself and its internal organization at a high level. These files are created as ASCII text files. A message protocol has been defined. This includes the medium (files), the syntax (NDAS Query Language—NQL), and semantics (program actions).

At a minimum, each communication includes:

1. sender ID,
2. receiver ID (may be broadcast or single recipient),
3. language (ASCII or binary),
4. ontology (defines the meaning of the syntactic expressions), and
5. file name (where message can be found).
6. history (names of all the agents which have played some role in giving rise to this message).

In general, an ontology is similar to a database schema and gives a specification of the objects, concepts and relationships that exist in the domain of interest. For NDAS, this has been defined by means of a semantic network. Concepts are represented by objects and relations between objects; these objects for the technical drawing analysis are called structures, and can be either structures (terminals) derived from the image, or model structures (nonterminals) which arise in the semantic model definition.

5.3 Constraint Handling

One of the aspects of engineering drawing analysis which requires an efficient mechanism is the determination of relations between the various entities in and extracted from an image. Pixels are related to text and graphics primitives, and primitives have relations like *near*, or *parallel*, etc.

Agent interactions may also be defined through a *goal graph*; this helps define the dependencies between agents and data, as well as allowing for higher-level agents to assign parts of the problem or determine the coherence of the progress of the independent agents. This new approach has been developed and is described in Chap. 4.

5.4 Semantic Networks and Agents

Higher-level models involve descriptions of the semantics of the drawings, and as such involve determining the relations between the primitives of the drawing and their meanings. We use graph models (semantic nets), which are explored through a grammatical paradigm. After the document is digitized, vectorized and

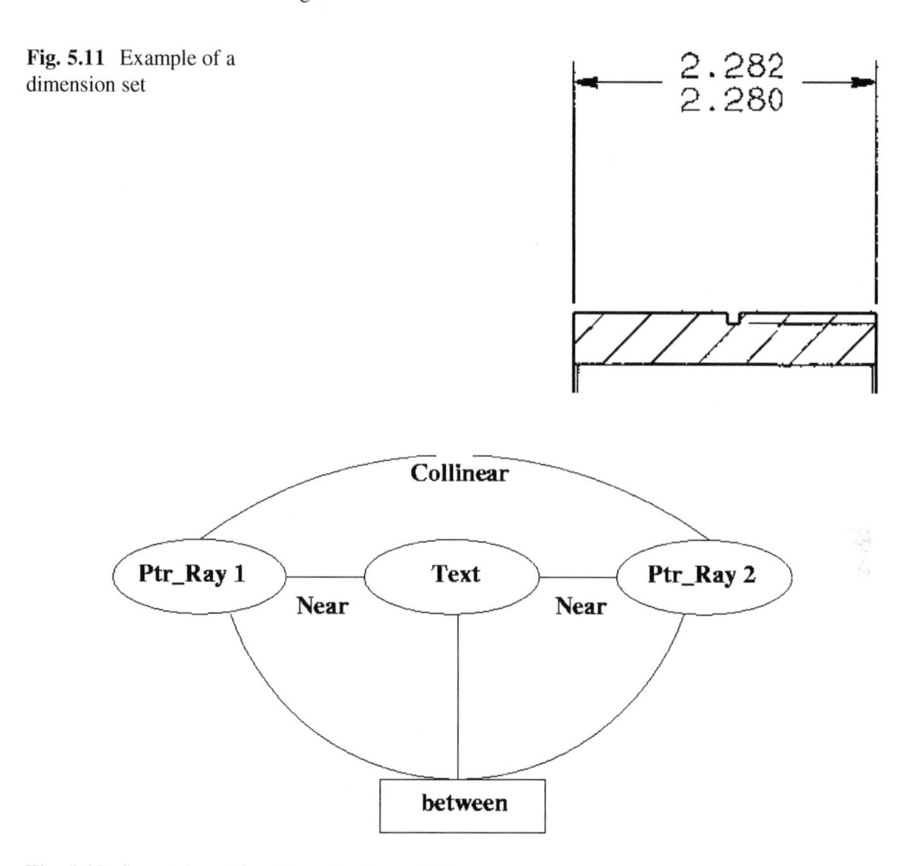

Fig. 5.11 Example of a dimension set

Fig. 5.12 Semantic net for dimension_Rays_Out

the connected components extracted, the result is a set of image primitives such as segments, arcs, arrows and text blocks. Based upon the relationships that exist between these primitives, a semantic net model is used to recognize important features like dimensions, annotations, and legends in the document. This is described in detail in Chap. 4. (Note: a semantic network may represent knowledge to be used to perform the low-level image analysis as well.) The relational structure of technical drawings and the analysis of digital images of such drawings are driven by a semantic network. This constitutes the high-level model. The image is analyzed for text, geometry (2D and perhaps 3D), regions, boundaries, and relations between the extracted objects. As an example higher-level structure, consider a *box_description* which is comprised of a *text* structure and a *box* structure, where the *text* is inside the *box*.

Figure 5.11 shows an instance of dimensioning taken from the scanned image of an actual technical drawing. A sample semantic net describing the association between various primitives is shown in Fig. 5.12.

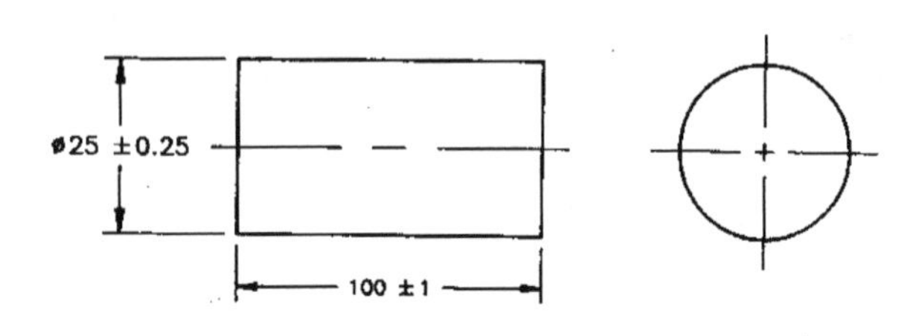

Fig. 5.13 Synthetic image that is analyzed

Table 5.1 Ground Structures and their count

Ground structures	Count
pointer_ray (pr1,. . . . pr4)	4
circle (c1)	1
text (t1,. . . ,t4)	2
boxes (box1)	1
line_segment (l1,. . . l21)	21

5.5 NDAS Experiments

We have tested the NDAS approach in a set of experiments in order to answer the following questions:

1. How good is the NDAS technical drawing analysis in absolute terms?
2. How good is NDAS relative to other methods?
3. How well do the complexity reduction techniques work?

We have strong answers to questions (1) and (3), but the response to (2) is more problematic.

5.5.1 *Ideal Analysis*

Figure 5.13 shows an engineering drawing which serves as our basic test image. It has a set of terminal symbols (e.g., **line_segment**, **pointer_rays**, **text**, **circle**) which form non-terminals of interest (e.g., *dimension_description*). Table 5.1 gives the ideal ground structures that are in the drawing and the possible number of vocabulary symbols that can be derived from those ground structures. These numbers will be used to evaluate the success of our methods in image analysis and complexity reduction.

Ground structures correspond to the terminal symbols in the grammar and are recovered by the image analysis agents. Figure 5.14 shows the drawing of Fig. 5.13 with ground structures labeled (only major line_segments are labeled):

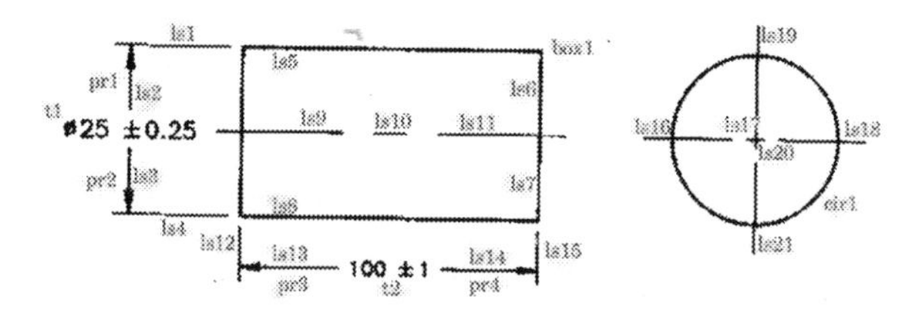

Fig. 5.14 Synthetic image with ground structures analyzed

Table 5.2 One ground structure

Image	Ideal results	True positive	False positive	False negative
Figure 5.15	12	12	109	0
Figure 5.16	5	4	2	1
Figure 5.17	6	5	0	1
Figure 5.18	10	7	0	3
Figure 5.19	8	5	3	3
Figure 5.20	444	364	0	80
Figure 5.21	25	25	0	0
Figure 5.22	6	6	0	0

If we apply the rules of the grammar G (See Appendix A) to the ground structures listed in Table 5.1, then the ideal count of vocabulary symbols is as listed in column 3 of Table 5.5. These counts provide the ground truth basis by which to evaluate the NDAS analysis.

5.5.2 *Image Analysis*

The image analysis agents have been applied to various images from our technical drawing image dataset. This dataset includes:

- images each containing instances of one ground structure
- images with combinations of ground structures, and
- images of general technical drawings with corresponding ground structures.

5.5.2.1 One Ground Structure Images

A summary of the results on images with one class of ground structure each comparing performance of the image analysis system to ideal performance is given in Table 5.2. These tests were run with a set of fixed thresholds. Figures 5.15–5.22 show the images and terminals obtained from them.

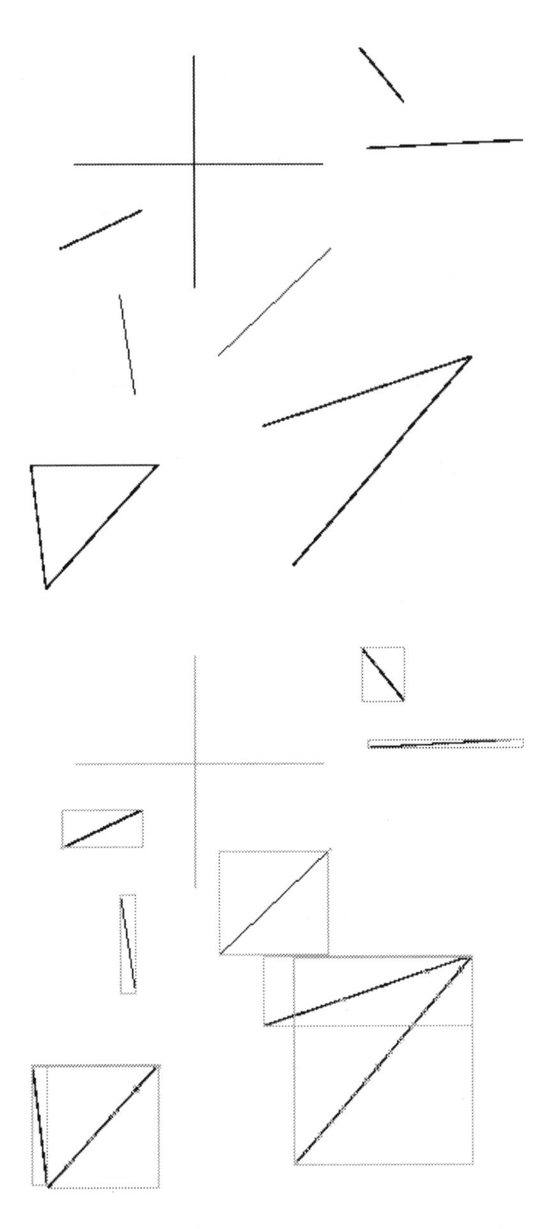

Fig. 5.15 Linesegments symbol instances

5.5.2.2 Combinations of Ground Structures

A summary of the results on combination of ground structures comparing performance of the image analysis system to ideal performance is given in Table 5.3. Figures 5.23 and 5.24 show the results.

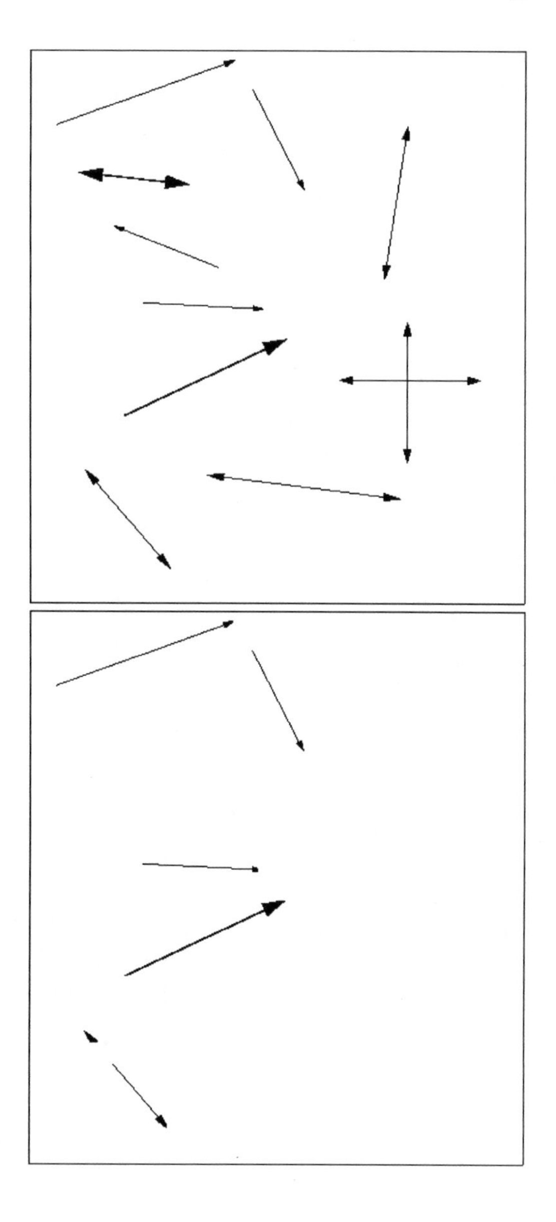

Fig. 5.16 PointerRays symbol instances

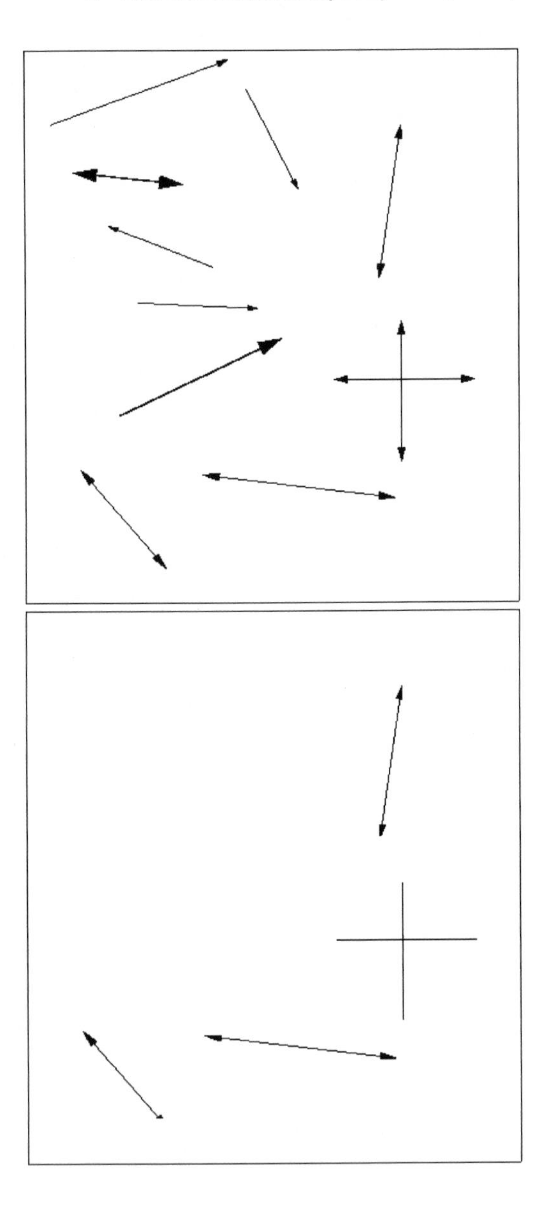

Fig. 5.17 PointerLines symbol instances

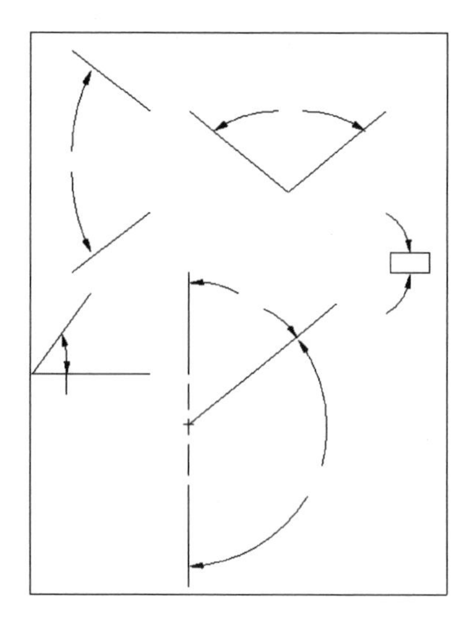
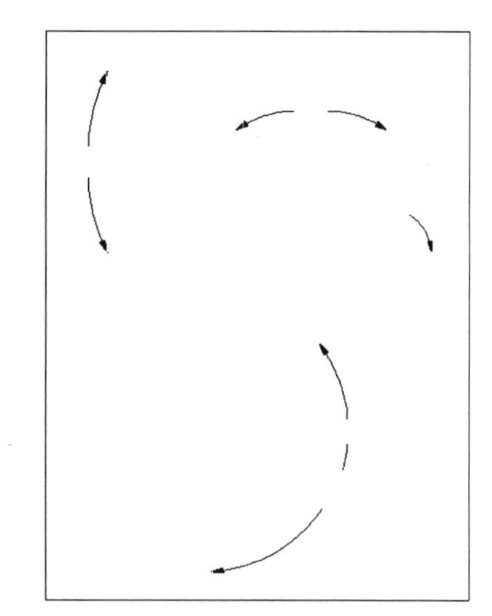

Fig. 5.18 ArcPointerRays symbol instances

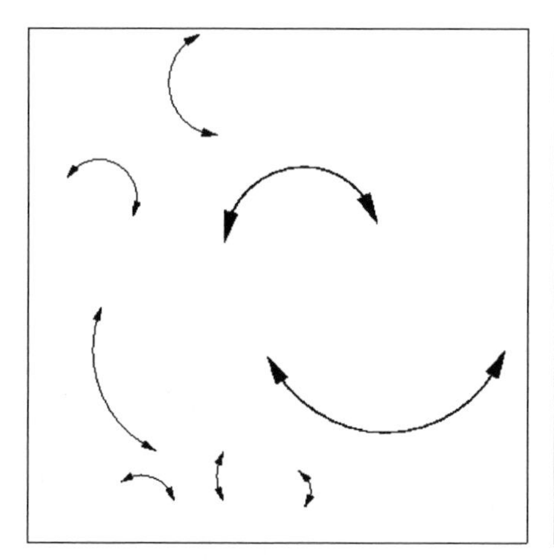
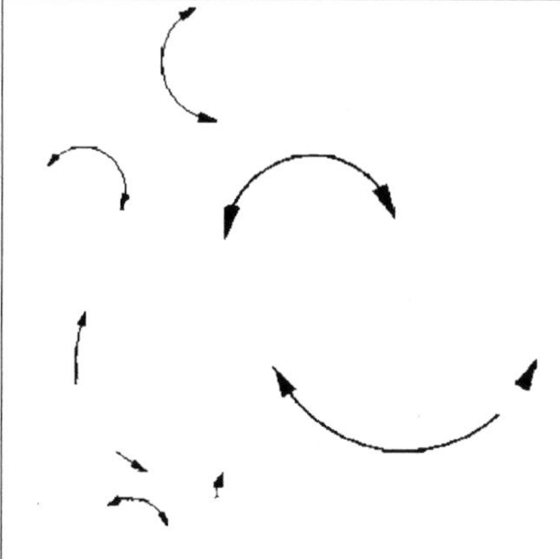

Fig. 5.19 ArcPointerLines symbol instances

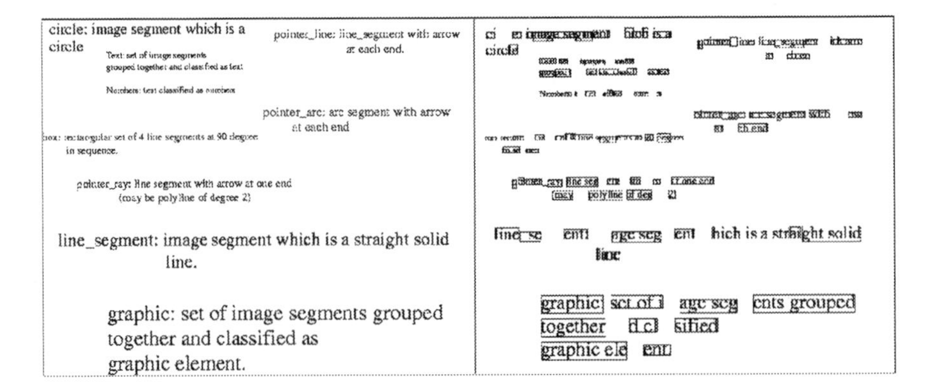

Fig. 5.20 Text symbol instances

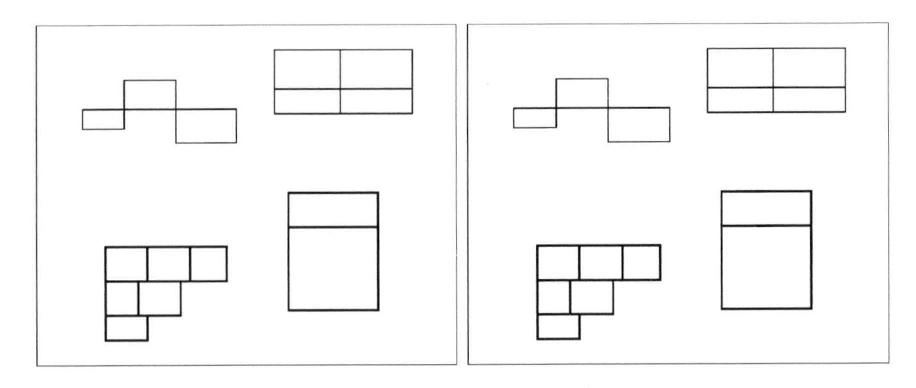

Fig. 5.21 Box symbol instances

5.5.2.3 General Technical Drawings

We describe the results of image processing algorithms on general technical drawings of Figs. 5.25 and 5.26 in this section. Figures 5.27 and 5.28 show the results. Table 5.4 gives a comparison of the performance of the image analysis system to ideal performance in detecting the ground structures.

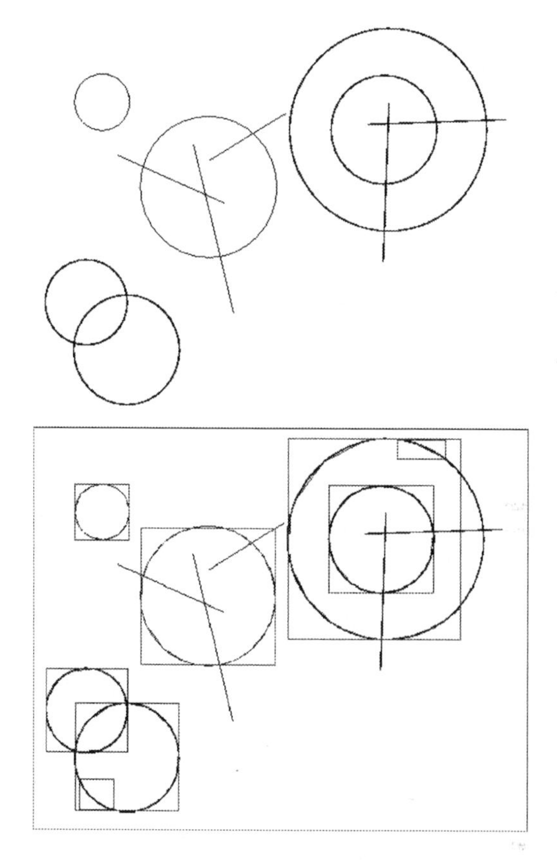

Fig. 5.22 Circle symbol instances

Table 5.3 Combination of ground structures

Image	Ideal results	True positive	False positive	False negative
Figure 5.23 (Circles)	3	1	0	2
Figure 5.23 (PointerRays)	1	1	0	0
Figure 5.24 (Circles)	3	2	0	1
Figure 5.24 (PointerLines)	1	1	1	0

5.5.3 Structural Analysis

We have chosen to separate out the structural analysis and to first evaluate it independently of the image analysis. (An overall analysis of the complete NDAS system is given in Sect. 5.5.5). To determine how well the structural analysis is performed, we have applied the structure-agent to the ideal sets of ground structures. This allows us to determine the actual number of symbols produced. *SR* refers to symbolic redundance and *SP* refers to symbolic pruning (see Sect. 4.4.1). Table 5.5 gives the symbolic redundancy for the vocabulary symbols in the grammar G (see

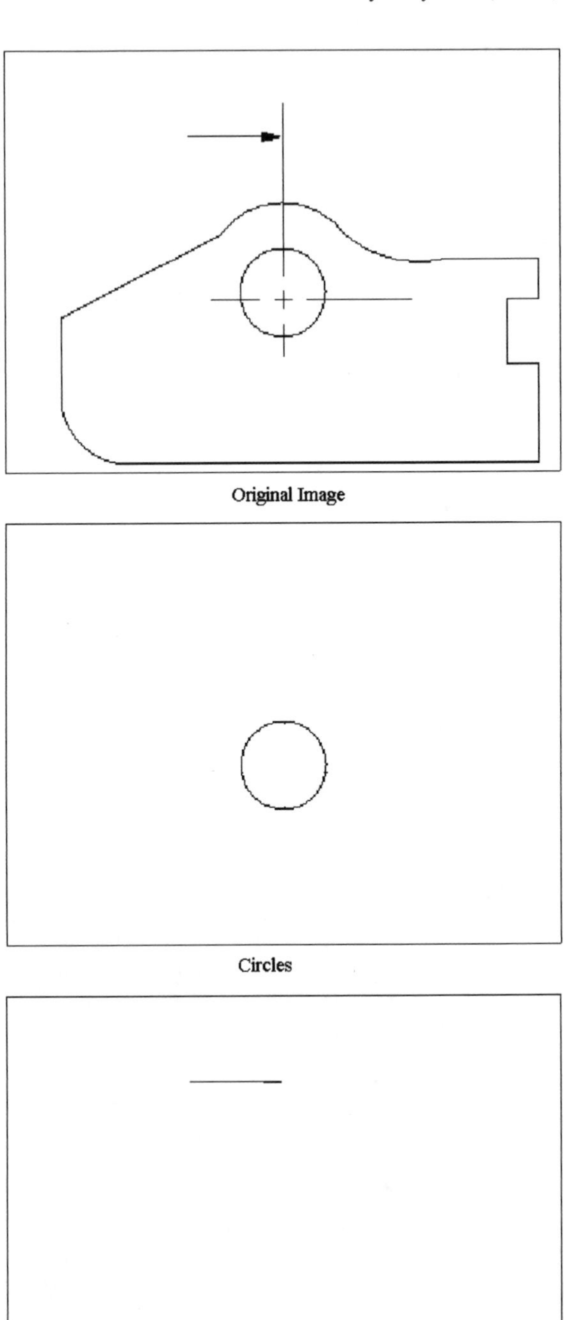

Fig. 5.23 Combination of ground structures 1

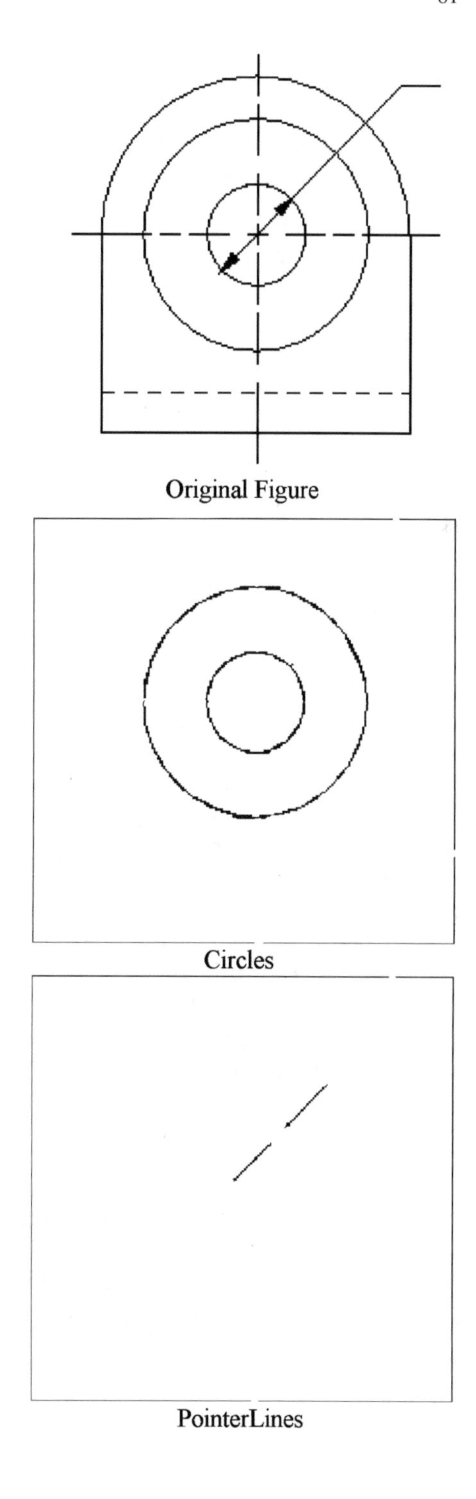

Fig. 5.24 Combination of ground structures 2

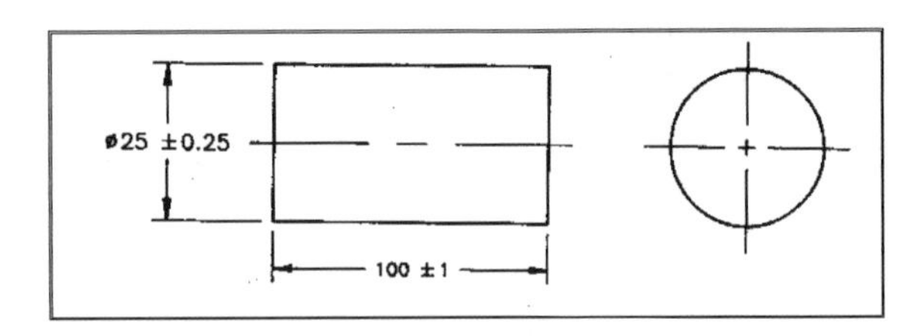

Fig. 5.25 Technical drawing 1

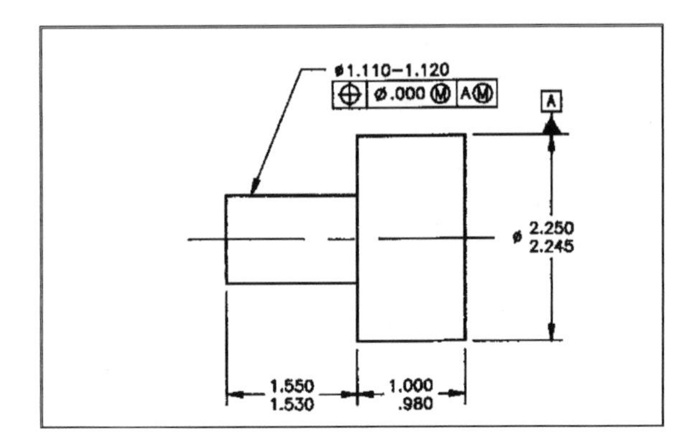

Fig. 5.26 Technical drawing 2

Appendix A) given the ideal ground structures. Table 5.6 gives the comparison between the system run with no pruning versus symbolic pruning.

The technical drawing in Fig. 5.25 has 2 ideal results as shown in Fig. 5.29. But the table shows 4 true positives and 0 false positives. The 4 true positives are due to repetition of the same result twice. Had symbolic pruning been applied, it would have resulted in only 2 true positives.

The technical drawing in Fig. 5.26 has only 4 ideal results (Figs. 5.30, 5.31, 5.34 and 5.37). But our structural analysis gives 8 results. In addition to the 4 ideal results, our analysis also detects four partially correct results (Figs. 5.32, 5.33, 5.35 and 5.36). These four *dimension_sets* are detected with the same **pointer_rays**, **line_segments** and **graphics** as the ideal ones but different texts. We term them *partially correct* since they are subsets of the ideal results.

5.5.4 Complete Image and Structural Analysis

We describe the complete analysis of Figs. 5.25 and 5.26 using the best set of thresholds for the image analysis and structural analysis agents thus producing ground structures and other nonterminal structures.

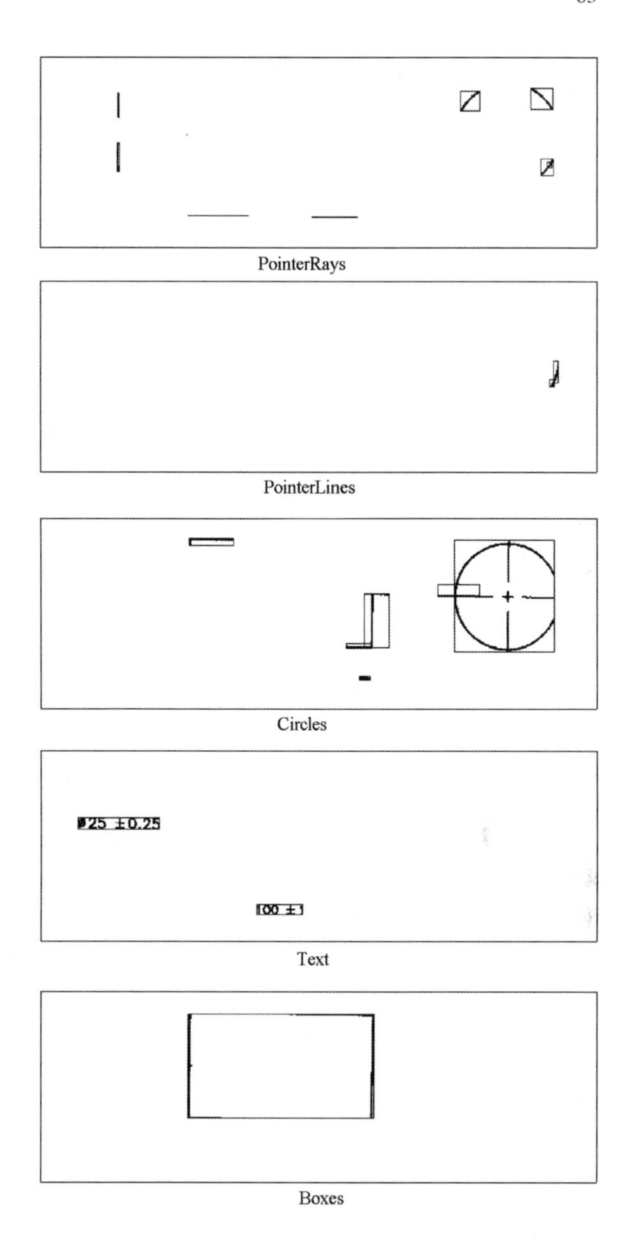

Fig. 5.27 Analysis of technical drawing 1

Fig. 5.28 Analysis of technical drawing 2

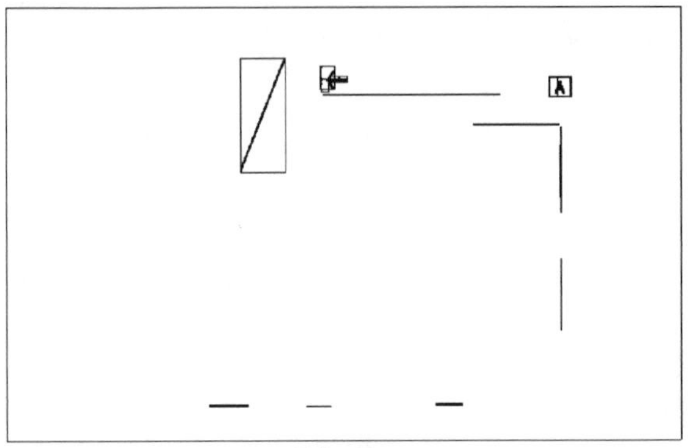

PointerRays

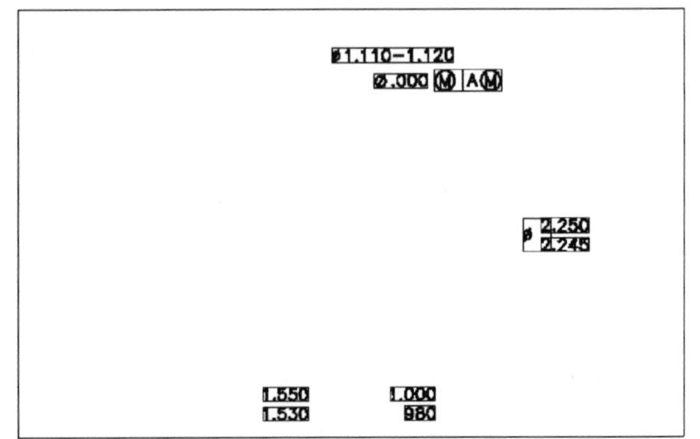

Text

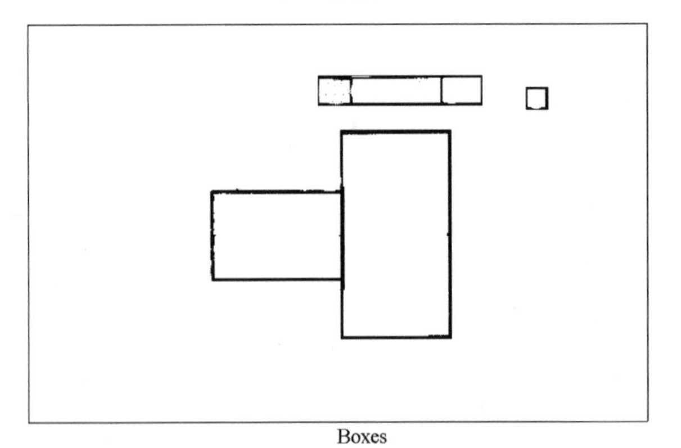

Boxes

Table 5.4 General technical drawings

Image	Ground structure	Ideal results	True positives	False positives	False negatives
Figure 5.27	*PointerRays*	4	4	4	0
Figure 5.27	*PointerLines*	0	0	2	0
Figure 5.27	*Circles*	1	1	5	0
Figure 5.27	*Boxes*	1	1	0	0
Figure 5.27	*Text*	2	2	0	0
Figure 5.28	*PointerRays*	7	6	4	1
Figure 5.28	*Boxes*	9	9	0	0
Figure 5.28	*Text*	11	9	1	2

Table 5.5 Symbolic redundancy for vocabulary symbols

Vocabulary symbol	SR(worst case)	SR(practice)+SP
line_segment	21	21
pointer_ray	4	4
text	2	2
circle	1	1
box	1	1
pointer_ray1	4	4
pointer_ray2	4	4
line_segment1	21	21
line_segment2	21	21
line_segment3	21	21
text1	2	2
text2	2	2
text_comb	2	0
text_final	4	2
symmetric_pointer_pair_in	12	0
symmetric_pointer_pair_out	12	2
dimension_rays_in	48	0
dimension_rays_out	48	2
dimension	96	2
dimension_set	40,320	2
pointer_ray_extn	84	0
check_sign	420	0
check_pair	1,680	0
dimension_description	>100,000	2
text_in_box	4	0
text_in_box1	4	0
text_in_box2	4	0
text_in_box3	4	0

(continued)

Table 5.5 (continued)

Vocabulary symbol	SR(worst case)	SR(practice)+SP
text_in_box4	4	0
one_datum_ref	0	0
datum_ref	0	0
datum_below_text	0	0
dashed_lines	7,980	3
dash_lines1	7,980	3
dash_lines2	7,980	3
circle_center_dim	>100,000	1

Table 5.6 Structural analysis

Image	Ideal results	True positives	False positives	Partially correct	%Detected	%Missed
Figure 5.29	2	4	0	0	100	0
Figures 5.30–5.37	4	4	0	4	100	0

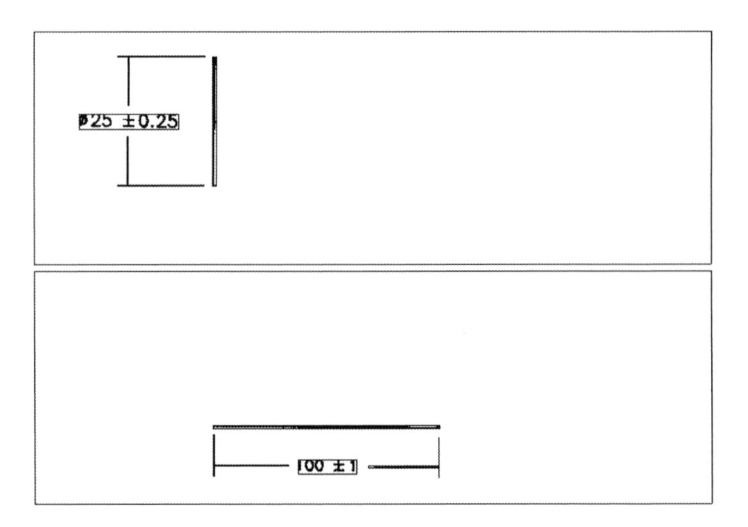

Fig. 5.29 Structural analysis of drawing 1

The analysis on Fig. 5.25 produces outputs shown in Figs. 5.38 and 5.39 which match the structural analysis output in Fig. 5.29.

Figure 5.40 is close to Fig. 5.31, Fig. 5.41 is close to Fig. 5.37 and Fig. 5.42 is close to Fig. 5.34. Figure 5.43 is erroneous, and one of the ideal outputs in the structural analysis (see Fig. 5.30) is not detected at all (Figs. 5.44–5.46). The rest of them are termed as *partially correct* results. A comparison of the ideal image analysis system with the above complete analysis by our system is given in Table 5.7. CA refers to complete analysis.

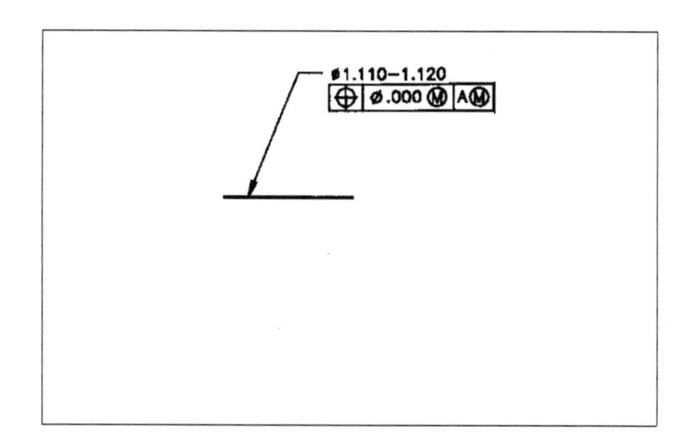

Fig. 5.30 Structural analysis result 2(a)

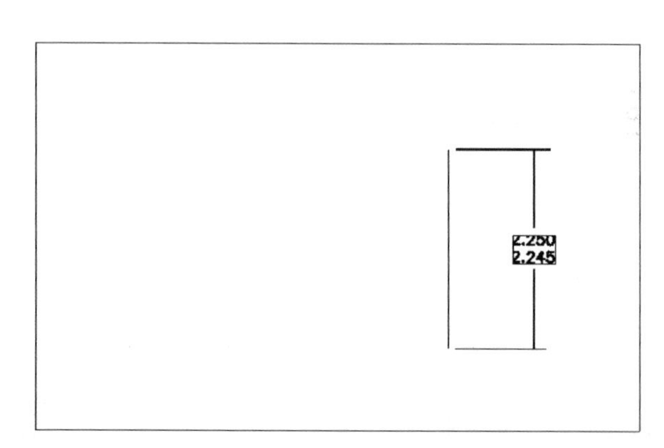

Fig. 5.31 Structural analysis result 2(b)

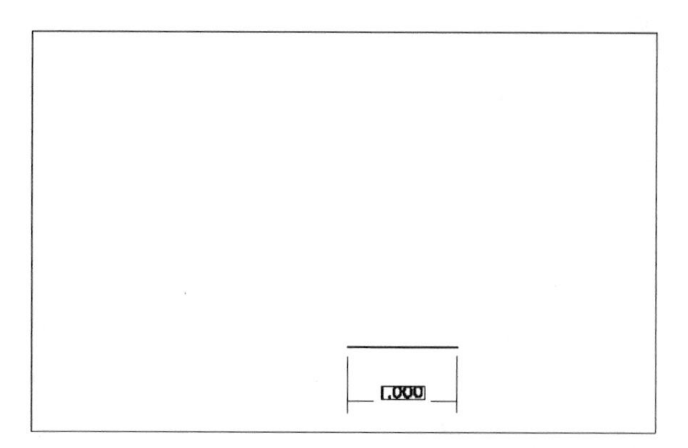

Fig. 5.32 Structural analysis result 2(c)

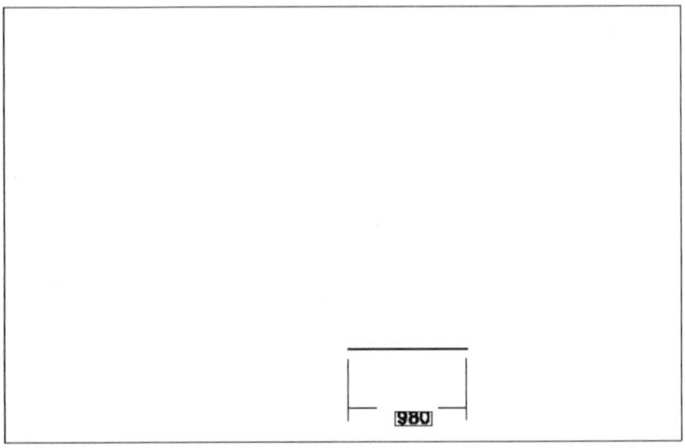

Fig. 5.33 Structural analysis result 2(d)

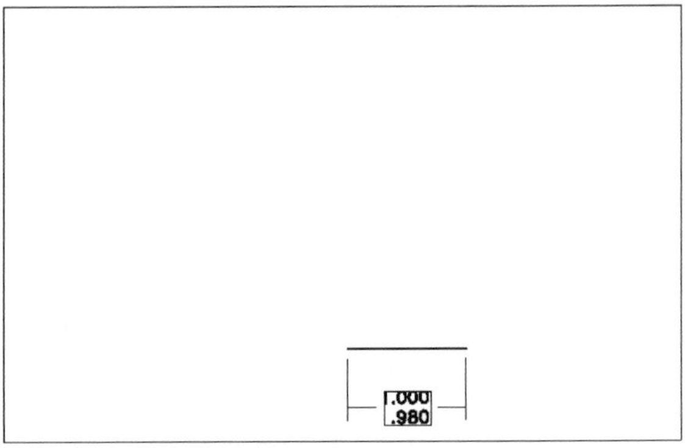

Fig. 5.34 Structural analysis result 2(e)

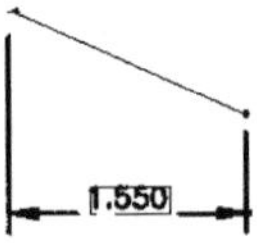

Fig. 5.35 Structural analysis result 2(f)

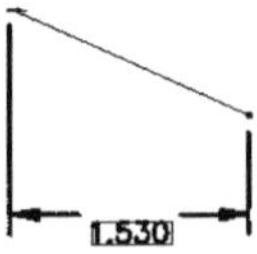

Fig. 5.36 Structural analysis result 2(g)

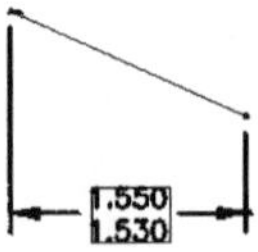

Fig. 5.37 Structural analysis result 2(h)

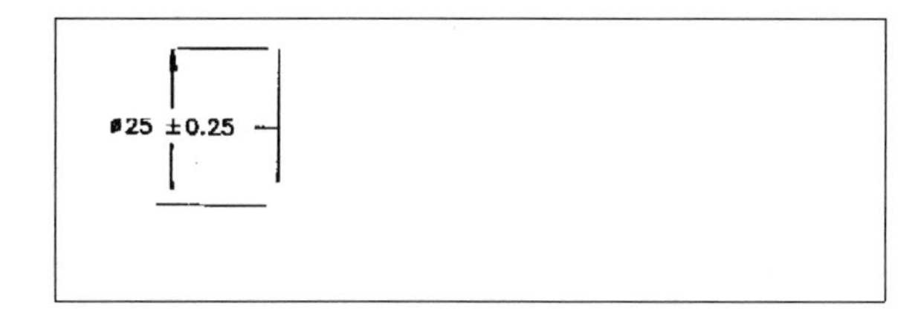

Fig. 5.38 Complete analysis result 1(a)

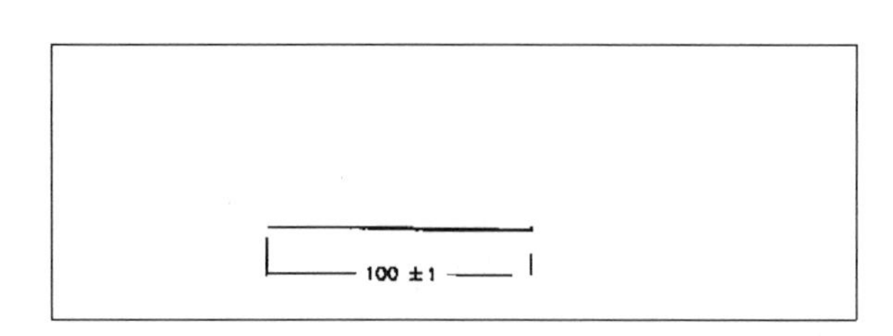

Fig. 5.39 Complete analysis result 1(b)

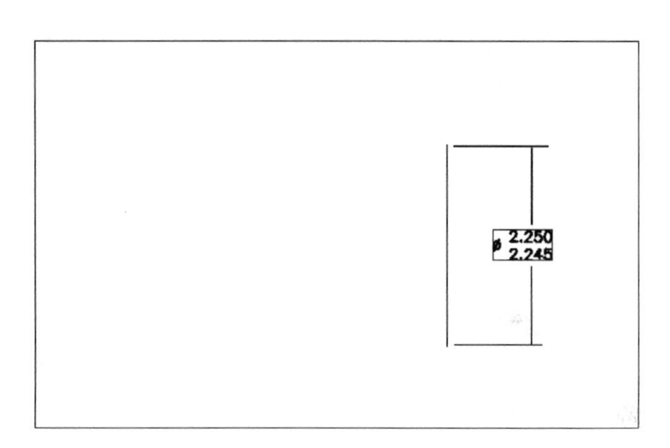

Fig. 5.40 Complete analysis result 2(b)

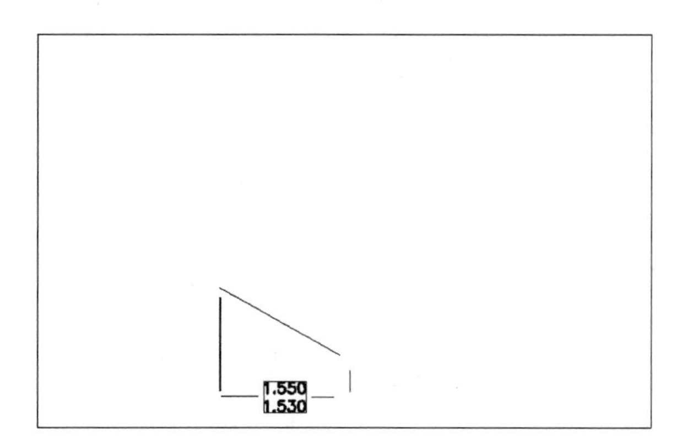

Fig. 5.41 Complete analysis result 2(d)

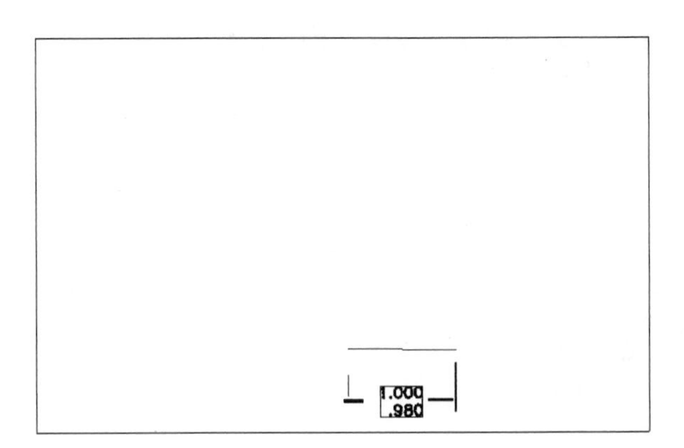

Fig. 5.42 Complete analysis result 2(g)

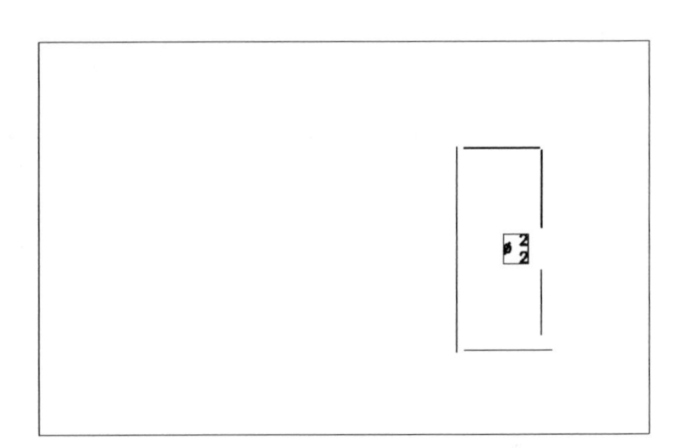

Fig. 5.43 Complete analysis result 2(a)

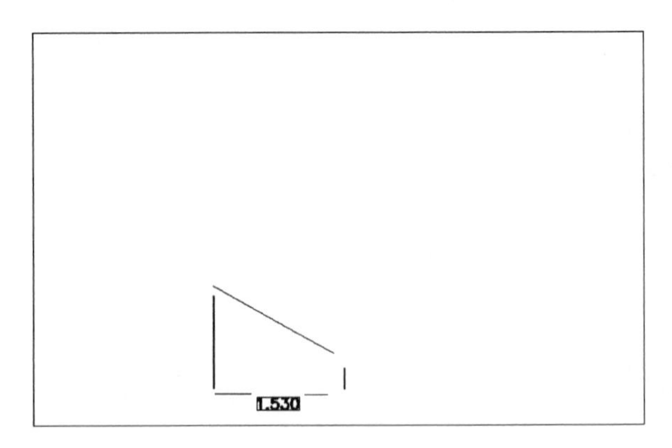

Fig. 5.44 Complete analysis result 2(c)

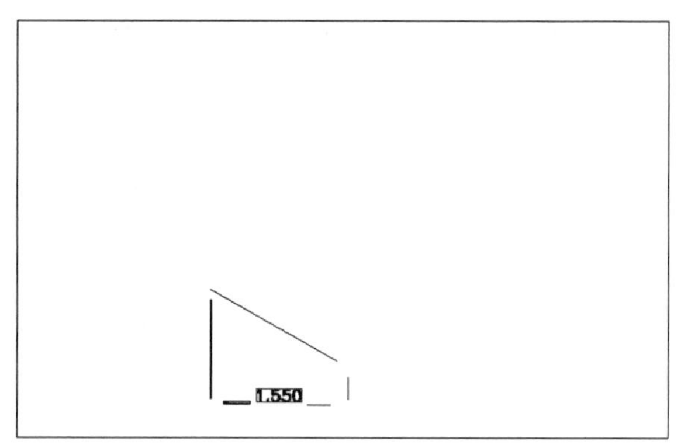

Fig. 5.45 Complete analysis result 2(e)

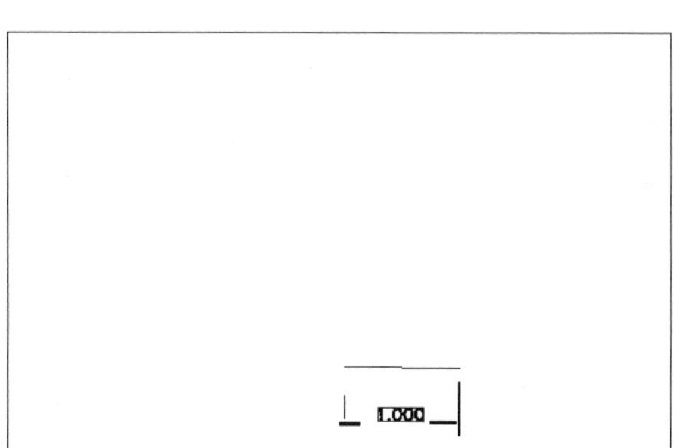

Fig. 5.46 Complete analysis result 2(f)

Table 5.7 Structural analysis for complete NDAS run

Image	Ideal results	True positives	False positives	Partially correct	%Detected	%Missed
CA 1	2	4	0	0	100	0
CA 2	4	6	2	6	75	0

5.5.5 Complete NDAS Analysis

The complete NDAS analysis consists of the application of both image analysis and structural analysis agents with multiple sets of thresholds.

5.5.5.1 Image Analysis

The result of the image analysis agents run with different sets of thresholds on Fig. 5.25 yields the cumulative result shown in Figs. 5.47–5.51. These results are then used by the structural analysis agents.

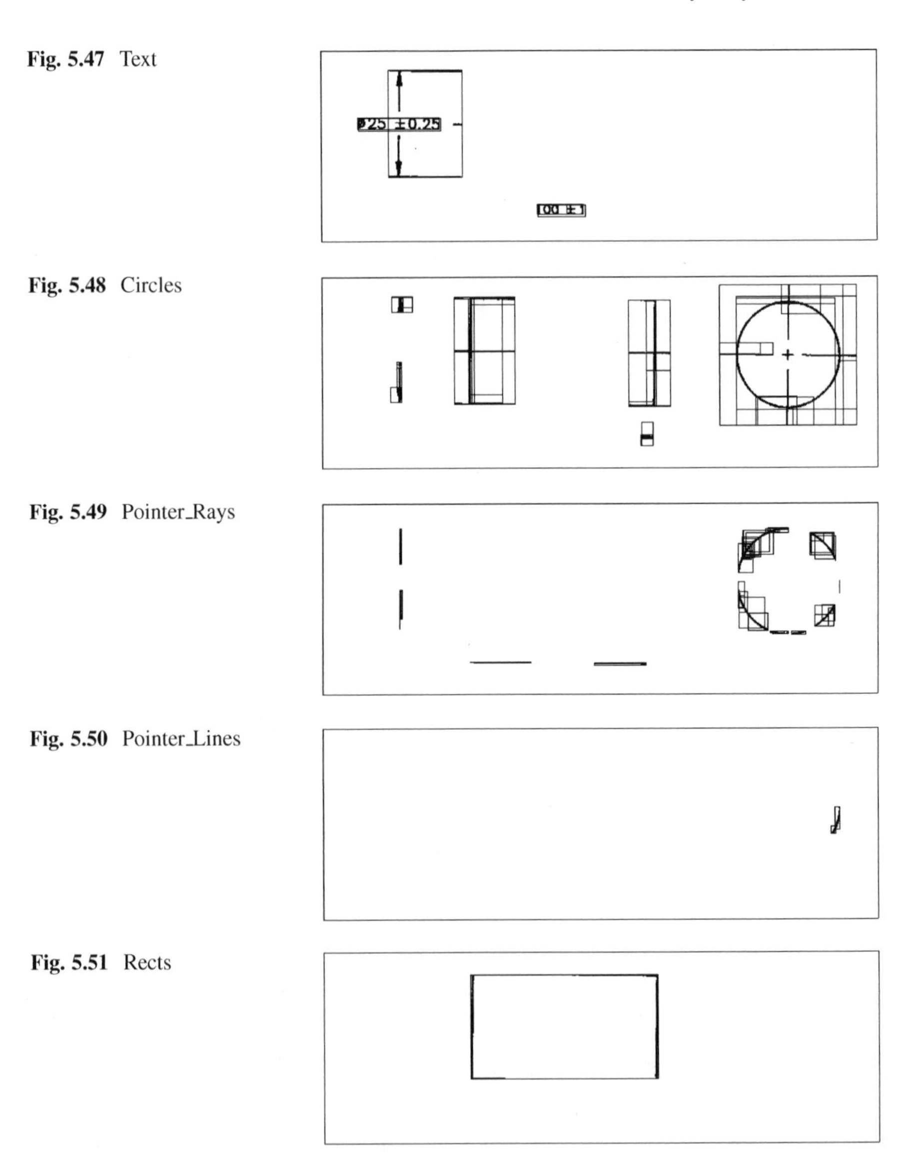

Fig. 5.47 Text

Fig. 5.48 Circles

Fig. 5.49 Pointer_Rays

Fig. 5.50 Pointer_Lines

Fig. 5.51 Rects

5.5.5.2 Output Reduction Methods

The result of every image-processing algorithm is subjected to further analysis by other agents, and these agents use various thresholds to produce more outputs and so on in an exponential manner. We have explored two ways of containing this growth.

1. As each result is produced, every agent casts a message to a Compare Agent whose job is to compare the current output with already existing outputs (of the same category) and broadcast a message asking other agents to neglect the

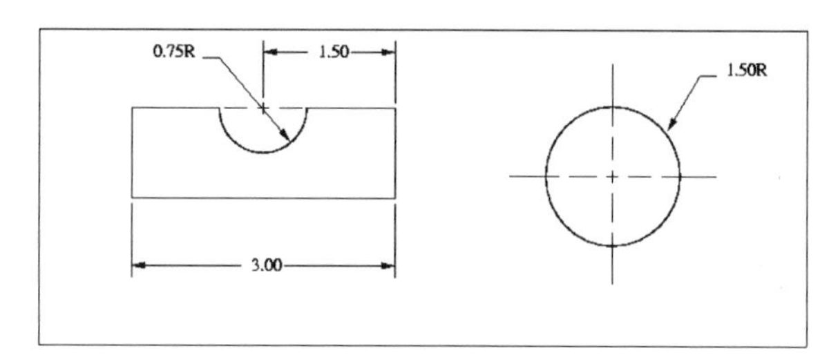

Fig. 5.52 Original image

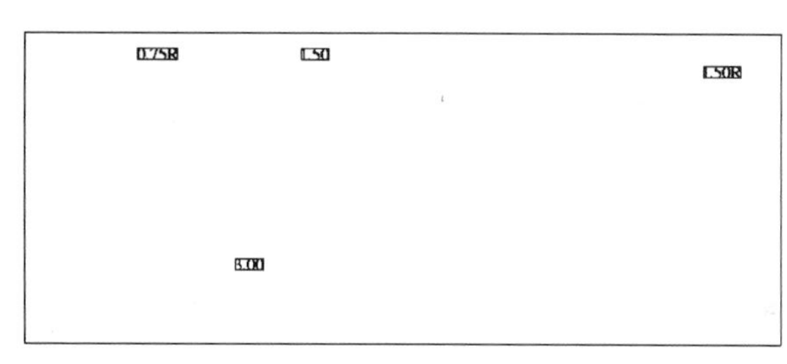

Fig. 5.53 Text detection 1

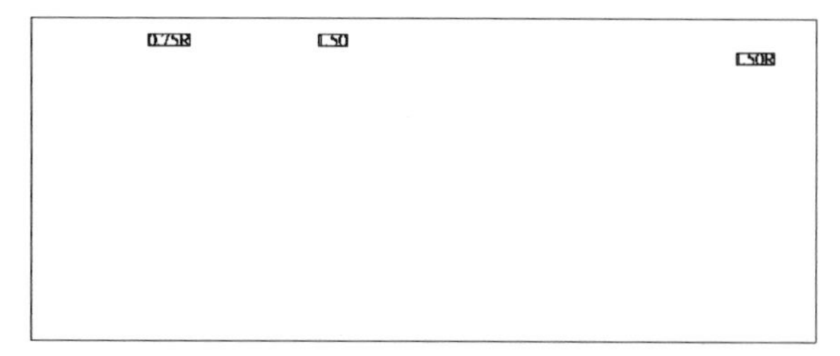

Fig. 5.54 Text detection 2

current output if either the output is erroneous (e.g., a matrix is empty, an image file is all zero, etc.) or if the output has already been produced.
Thus, whenever an agent comes across such a message (denoting that a file is to be ignored), the agent stores this filename in its state and thereby avoids working on it. If it already started working on it, it doesn't cast the usual output message on completion. If it had finished working on it and already cast the output message on completion, it broadcasts another message to all agents asking them to ignore that output.

2. As and when results (of a category) are produced, the user may be asked to eliminate a few outputs by viewing them (i.e., interactive empirical pruning). This can reduce the number of outputs to a great extent.
For example, a text detection algorithm was applied to Fig. 5.52 with various parameters. Four different outputs were produced among which Figs. 5.53 and 5.54 are good since they have no extraneous outputs. Instead of proceeding with all the sets of outputs, user interaction can be used to establish good thresholds and eliminate poor results thereby reducing the search space. Thus, the user might pick either Figs. 5.53 and 5.54, or just Fig. 5.53 and eliminate the rest (Figs. 5.55 and 5.56).

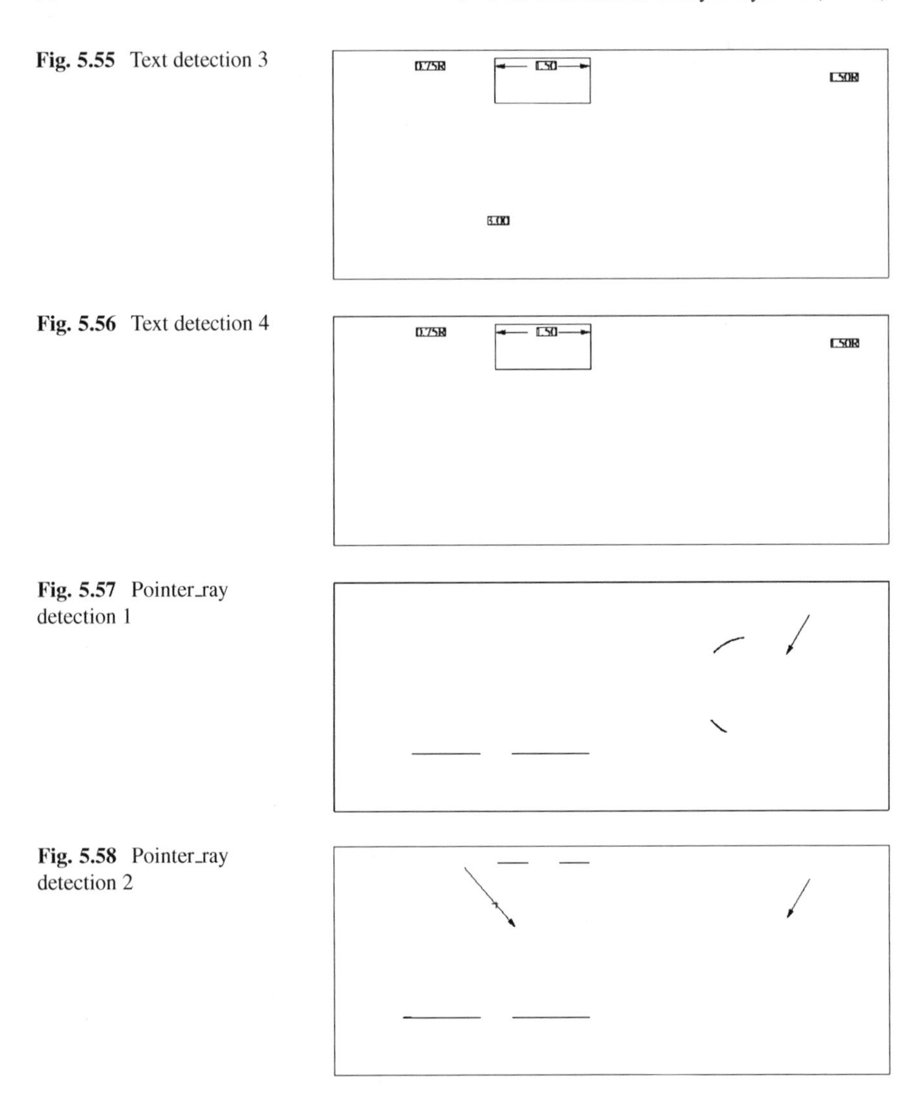

Fig. 5.55 Text detection 3

Fig. 5.56 Text detection 4

Fig. 5.57 Pointer_ray detection 1

Fig. 5.58 Pointer_ray detection 2

Figures 5.57–5.59 are the three different outputs of a pointer_ray detection algorithm obtained by using two different sets of thresholds. All the **pointer_rays** in Fig. 5.58 match the ideal output while Figs. 5.57 and 5.59 contain extraneous outputs in them. Hence, the user would benefit from choosing Fig. 5.58 and eliminating the others.

Finally, the above two methods can be put together to decrease the search space to a greater extent.

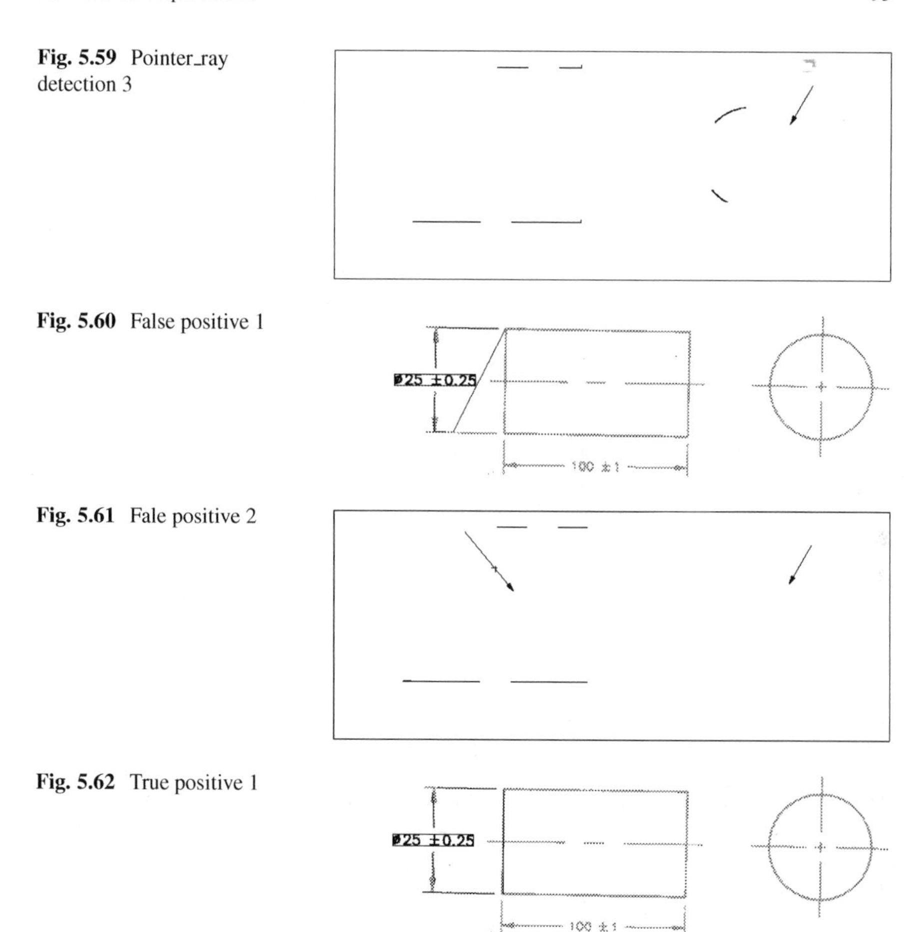

Fig. 5.59 Pointer_ray detection 3

Fig. 5.60 False positive 1

Fig. 5.61 Fale positive 2

Fig. 5.62 True positive 1

5.5.5.3 False Positives Analysis

The structural analysis agents operate on the results of image analysis agents shown in Sect. 5.5.5.1 to produce the highest_level desired vocabulary symbol: *dimension_description*. Figures 5.60 and 5.61 show the false positives and Fig. 5.62 shows one of the true positives obtained by the structural analysis on the results shown in Sect. 5.5.5.1.

The reason for false positives is the use of a set of thresholds. Since the best threshold for each of the image/structural analysis agents is not known to us a priori, it is difficult to achieve the highest true positive and lowest false positive percentages using just one set of thresholds. Hence, it is imperative to use different sets of thresholds. This results in many false positives. Empirical pruning reduces false positives to an extent, but not completely. The solution lies in improving both the image analysis algorithms and the rules of the grammar.

Our system provides a window through which both good and bad results enter. This helps later in the structural analysis phase to experiment with the full range of outputs. Had we employed just a single set of thresholds, we wouldn't know if all the *dimension_descriptions* would be detected.

5.5.5.4 False Negatives Analysis

False negatives can occur in two places in our NDAS system. They can occur in the image analysis phase if certain terminals go undetected for any value of threshold used in the analysis. The only way to prevent this is by improving the image analysis algorithms. These errors are carried over to the structural analysis phase and result in some *dimension_descriptions* not being detected because, one or more of the structures (terminals) comprising the *dimension_descriptions* (or parts of nonterminals in the *dimension_description*) were undetected in the image analysis phase.

False negatives can also occur if the rules are erroneous. Since we have had no false negatives due to our grammar G, we have not experimented with better rules.

5.5.6 Complexity Reduction Performance

The results obtained in the analysis of technical drawings were achieved by allowing multiple hypotheses to be generated in parallel and taking them all the way to successful interpretations. NDAS explores a large percentage of the search space and thus another goal of ours is to find ways to reduce the complexity of this search.

Section 4.4 describes the basis for our approach to achieve this. Here we present the results of applying our methods to the technical drawing analysis problem.

5.5.6.1 Symbolic Pruning

Given the drawing in Fig. 5.25, running the complete NDAS analysis produces a lot more symbols than those produced by the ideal analysis (Table 5.5). Table 5.8 shows the comparison of SR (worst case) with SR (in practice) with symbolic pruning.

5.5.6.2 Empirical Pruning

Grammar G uses many constraint relations such as *near*, *touches*, *parallel*, etc., that use parameters like *ANGLE_MAX* and *NEAR_LEN*. A specific threshold for these parameters may or may not yield all the desired results. For example, if the distance between the end of a **pointer_ray** and a **text** in an image is 25 pixels, a value of 15 for the *NEAR_LEN* parameter might not detect the **pointer_ray** and **text** pair.

Table 5.8 Symbolic redundancy for vocabulary symbols

Vocabulary symbol	SR(worst case)	SR(in practice) + SP
line_segment	326	326
pointer_ray	47	47
pointer_line	2	0
text	7	7
circle	42	42
box	8	8
pointer_ray1	47	47
pointer_ray2	47	47
line_segment1	326	326
line_segment2	326	326
line_segment3	326	326
text1	7	7
text2	7	7
text_comb	42	2
text_final	49	7
symmetric_pointer_pair_in	2,162	0
symmetric_pointer_pair_out	2,162	7
dimension_rays_in	>100,000	0
dimension_rays_out	>100,000	10
dimension	>100,000	10
dimension_set	>100,000	20
pointer_ray_extn	15,322	1
pointer_line_extn	652	2
pointer_line_extn_in_circle	27,384	0
check_sign	>100,000	256
check_pair	>100,000	0
dimension_description	>100,000	12
text_in_box	392	0
text_in_box1	392	0
text_in_box2	392	0
text_in_box3	392	0
text_in_box4	392	0
one_datum_ref	>100,000	0
two_datum_ref	>100,000	0
datum_ref	>100,000	0
datum_below_text	>100,000	0
dashed_lines	>100,000	13
dash_lines1	>100,000	13
dash_lines2	>100,000	13
circle_center_dim	>100,000	12

Hence, there is a requirement of multiple thresholds for the parameters. This in turn produces a large set of results and it becomes necessary to choose a set of thresholds which satisfies two criteria. The more important of these criteria is given first:

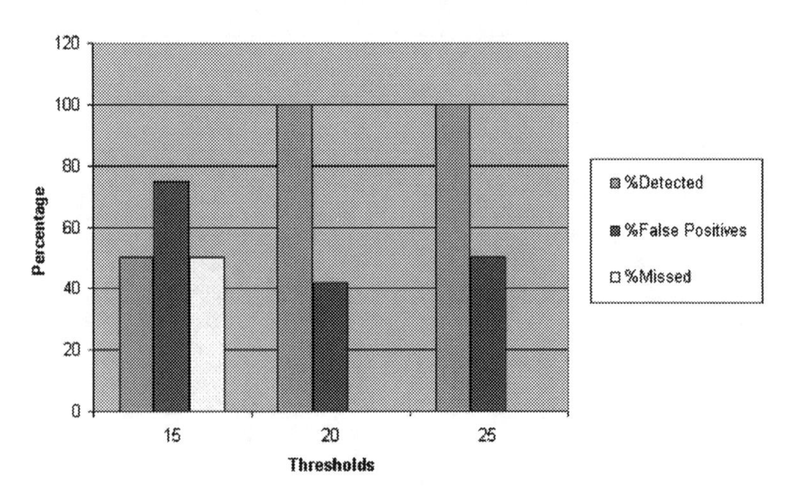

Fig. 5.63 Output of empirical pruning

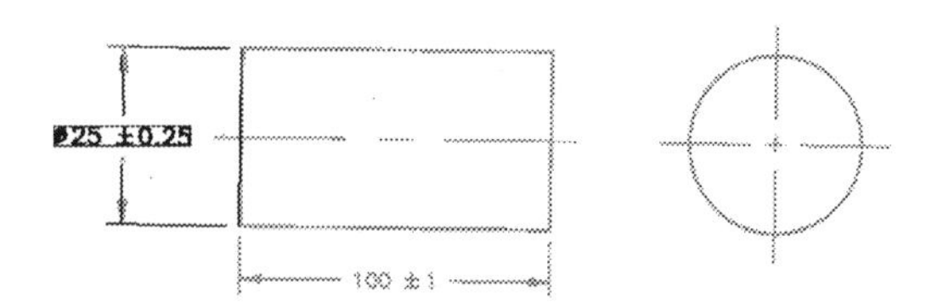

Fig. 5.64 Empirical output 1

1. highest detected percentage (= lowest missed percentage),
2. lowest false negative percentage.

We give preference over lower misses to lower false positives because our goal is to detect all the *dimension_descriptions* possible without missing any, even if the false positive percentage is high, rather than settling for fewer *dimension_descriptions* detected with fewer false positives.

User intervention is sought to find the best threshold. This is achieved by making the user view each result and reject the ones that are wrong.

Empirical pruning was applied on Fig. 5.25 with varying threshold values for *NEAR_LEN*. Figure 5.63 shows the percent detected, percent false positives and percent missed results using thresholds of 15, 20 and 25. A threshold value of 20 produces the best result.

Empirical pruning also gives a chance for the user to decide which result is best. For instance, Figs. 5.64 and 5.65 both depict the same *dimension_description* but the **pointer_ray** in Fig. 5.64 is closer to the ideal output than the other. The user might reject or accept the *dimension_description* produced in Fig. 5.65 accordingly.

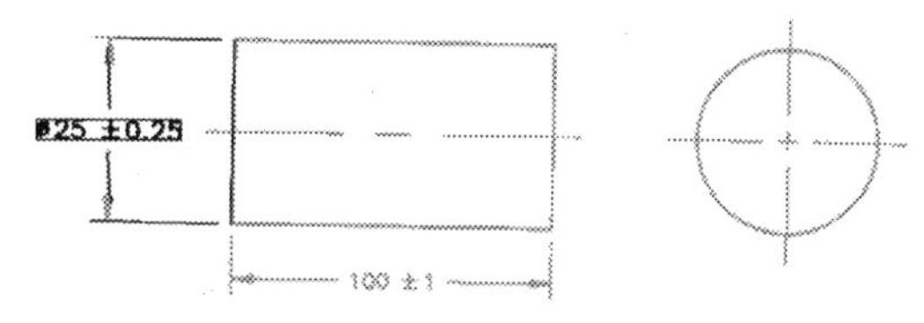

Fig. 5.65 Empirical output 2

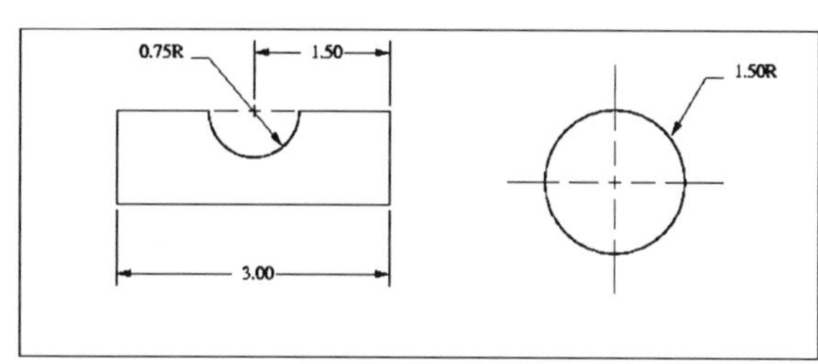

Fig. 5.66 Engineering drawing 1

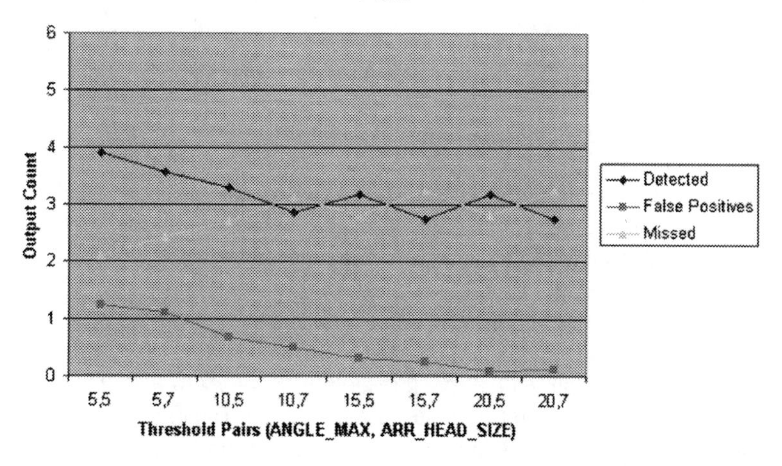

Fig. 5.67 Counts of detected, false positives and missed PointerRays vs threshold pairs

5.5.6.3 Threshold Sensitivity Analysis

During the image analysis stage of the NDAS system, different thresholds are used for parameters to produce different sets of outputs since it is not known which threshold will produce the best results on every image. In such a scenario, it is imperative to understand which threshold gives good results. The image analysis agent for determining **pointer_rays** was run on Fig. 5.66, and the results were compared to the ideal results. Figure 5.67 shows the graph of average count of detected **pointer_rays**, false positives and missed **pointer_rays** taken over a range of 186 outputs versus various threshold-pair values. The first threshold, *ANGLE_MAX*, gives the measure of maximum angle for parallelism, and the second,

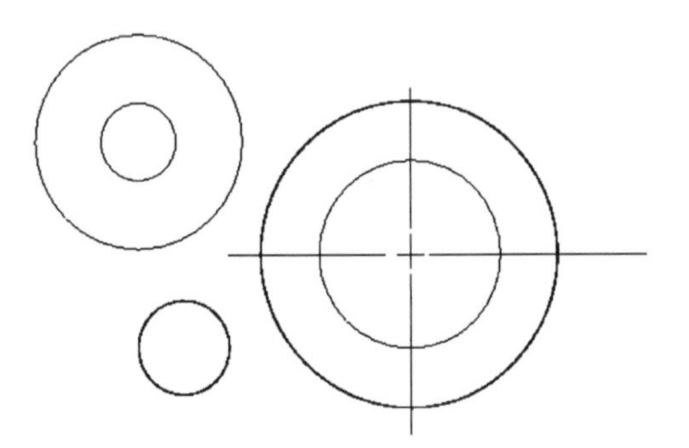

Fig. 5.68 Engineering drawing 2

ARR_HEAD_SIZE, gives the measure of pointer head size. The ideal number of **pointer_rays** is six. It can be inferred that both the count of detected and false positive **pointer_rays** seem to decrease with increasing values of *ANGLE_MAX* but for the same value of *ANGLE_MAX*, there is a greater count of detected, lower count of false positives and lower count of missed **pointer_rays** for a threshold of *ARR_HEAD_SIZE* equal to 5 rather than 7 (with constant threshold of *ANGLE_MAX*). It would be best for the user to choose the threshold pair (*5,5*) for the analysis even though the false positive count is the greatest for this threshold-pair because this is the only combination of thresholds that gives the maximum detected and minimum missed counts. The user could rely on the structural analysis phase to eliminate the incorrect **pointer_rays** (the false positives).

The image analysis agent for determining **circles** was run on Fig. 5.68, and the results were compared to the ideal results. Figure 5.69 shows the graph of percent detected versus various threshold-pairs used in the image analysis code. The first parameter, *NUM_BINS*, gives the number of bins in histograms used in the analysis and the second, *ANGLE_MAX*, gives the measure of maximum angle for parallelism. It can be seen from the graph that the threshold of 5.0 for *ANGLE_MAX* gives the maximum %detected for both the values of *NUM_BINS*. The user could choose the threshold pair of (*25,5.0*), to yield greater correctness in circle detection.

5.5.6.4 Precision and Performance Analysis

In our context, precision and performance analysis are determined on the final outcome after both image and structural analysis.

- *Precision*: Denotes how closely matched the *dimension_description* outputs (correct ones) produced by our agent-system are with respect to the ideal *dimension_description*. We study the closeness between the values of sub-structures of the *dimension_description* to the ideal ones in the image space.

  The structural agent acting on various outputs has parameters (thresholds) which control the precision of the final results. The two parameters that are

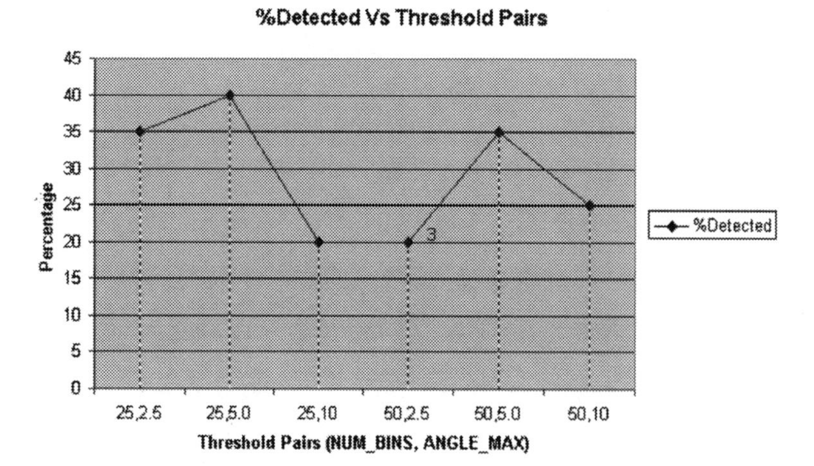

Fig. 5.69 Percent detected vs. threshold pairs

populated with different thresholds on different runs are *ANGLE_MAX* and *NEAR_LEN*. We employ thresholds of 10 degrees and 15 degrees for *ANGLE_MAX* (i.e., the structures comprising the *dimension_descriptions* may vary from the ideal structures by a maximum of 15 degrees of angle parallelism). The thresholds that we employ for *NEAR_LEN* range from 15 to 25. Hence, the final outputs may be said to vary from the ideal ones by up to a 25 pixel distance.
As described in Sect. 5.5.6.2, it is left to the user to decide whether or not some results be termed as false positives depending upon how precise they are. Figures 5.64 and 5.65 show such an output pair.

- *Performance analysis*: We propose three metrics for performance analysis:

  1. *Space*: The total number of symbols produced by the analysis.
  2. *Time*: The total time required to produce the analysis.
  3. *Detected, Missed and False Positives percentage* : The percentage of detected, missed and false positives of the structural analysis.

Figure 5.70 is the test image we presented in Chap. 4. The complete analysis on this image yielded the results shown in Figs. 5.71–5.73. The fourth *dimension_description*, is partially detected as shown in Fig. 5.74 (text inside the boxes aren't fully detected).

Table 5.9 shows the time taken by complete and reduction image analysis on images in Figs. 5.25 and 5.70. By reduction analysis, we mean the analysis incorporating the output reduction methods specified in Sect. 5.5.5.2. As can be seen, employing the reduction methods greatly reduces the run time. The analysis of all the drawings were done on an Athlon Processor with 1,499 MHz running the Linux operating system.

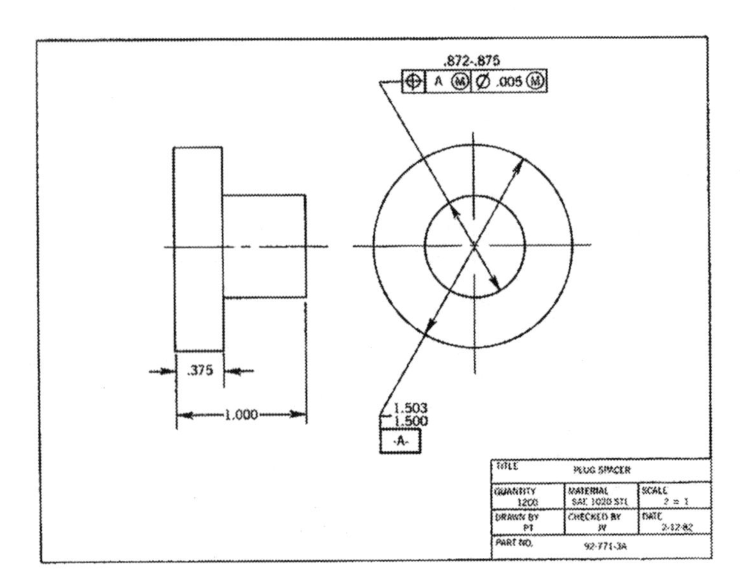

Fig. 5.70 Detailed engineering drawing

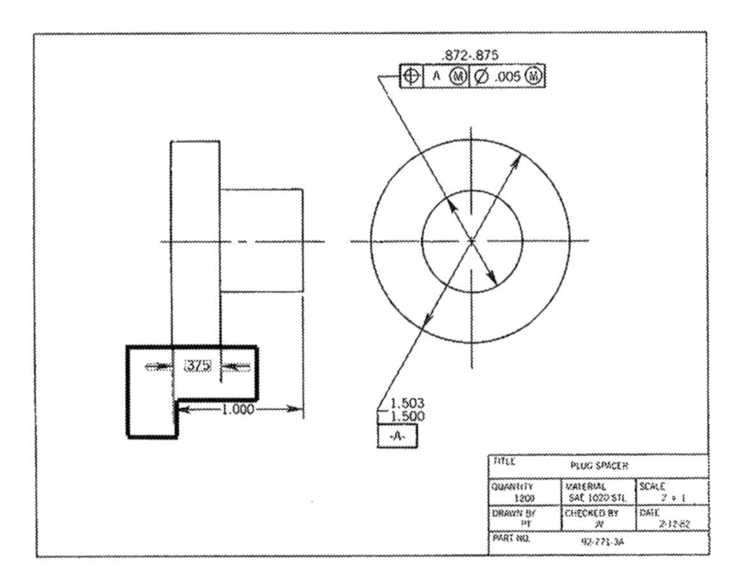

Fig. 5.71 Dimension 1

Figures 5.75 and 5.76 give the comparison of various factors on performance of the system on complete and reduction analysis with and without symbolic and empirical pruning on Figs. 5.25 and 5.70.

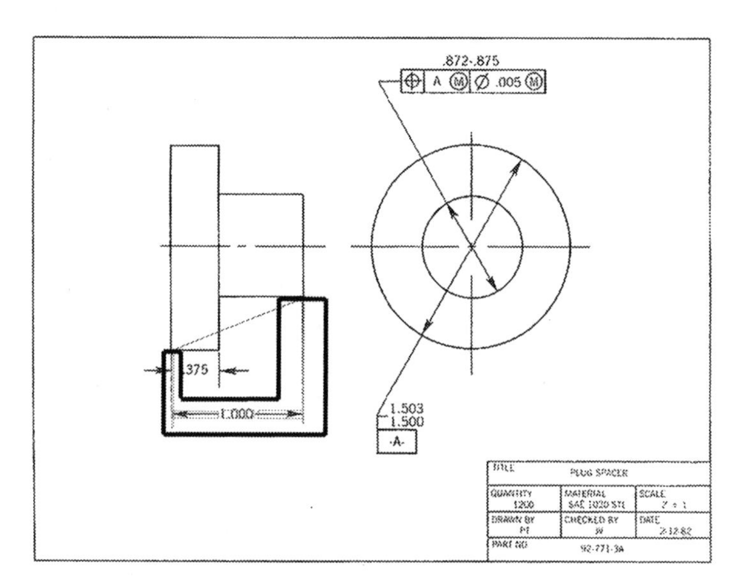

Fig. 5.72 Dimension 2

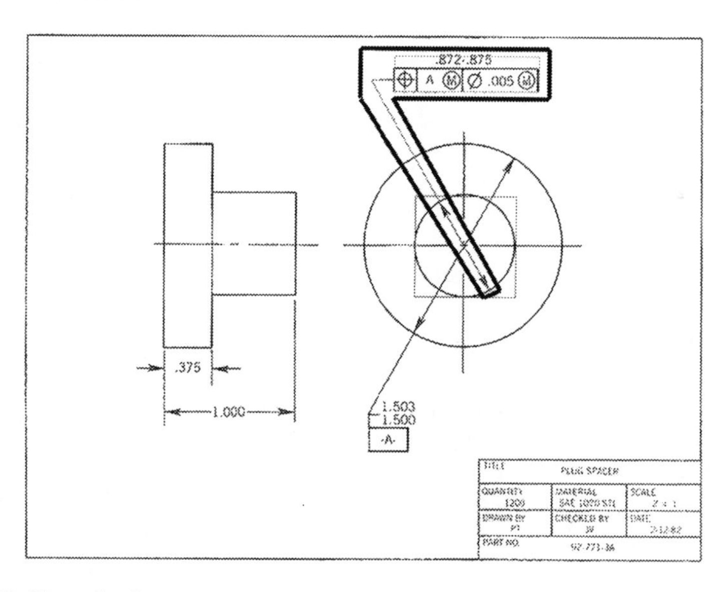

Fig. 5.73 Dimension 3

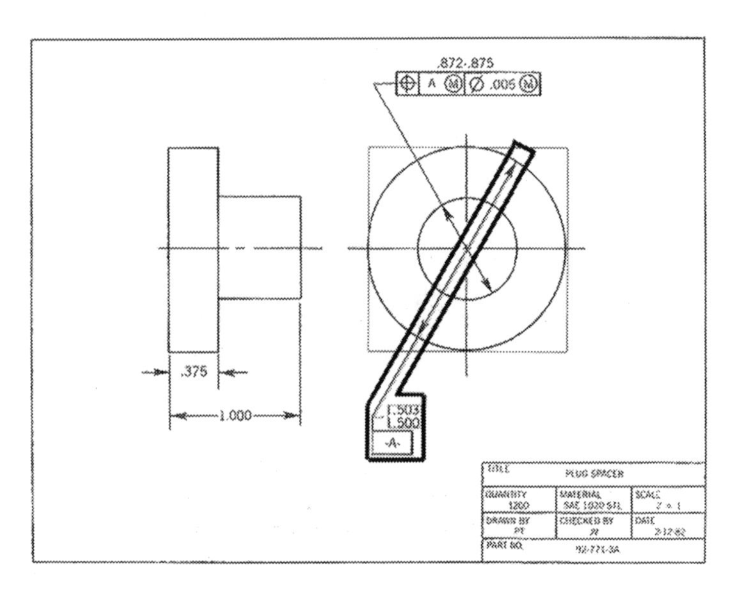

Fig. 5.74 Dimension 4

Table 5.9 Image analysis type and time metric

Figure	Analysis type	Time taken
Figure 5.25	Complete image analysis	15 h
Figure 5.25	Reduction image analysis	<2 h
Figure 5.70	Complete image analysis	49 h
Figure 5.70	Reduction image analysis	<9 h

5.5.7 *Higher Level Analysis*

5.5.7.1 Difference Between Complete and Reduction Analysis

Our primary goal is to find all ideal *dimension_descriptions* in the image (high percent detected and low percent missed), and, if possible, to reduce the number of spurious results (false positives).

Sometimes, the performance of reduction analysis in percent positives and percent false positives might be higher than those of the complete analysis. But this doesn't increase the percent detected or decrease the percent missed results. This is due to the fact that, as mentioned in Sect. 5.5.5.2, the second method of reducing outputs from the image analysis is through user intervention. If the user doesn't choose the correct structures (terminals), it might give rise to poor percent detected. Also, the reduction analysis in practice uses only a subset of the terminals used by a complete analysis. So, the percent missed of reduction analysis can never be lower than the percent missed of the complete analysis.

	Multiple Thresholds	*Empirical Pruning*	*Symbolic Pruning (for 1 vocabulary symbol)*	*Symbolic and Empirical Pruning (for 1 vocabulary symbol)*
Total no of dimension_descriptions: 2				
Complete Analysis				
Total no of symbols	8380	1458	57 (else, 114)	9 (else, 18)
Time taken to run	8 mins	<= 8 mins		
No of terminals	394	<= 394	< 394	< 394
% True positives	44.4	58.3		
% False positives	55.6	41.7		
% Damaged	0	0		
% Detected	83.3	100		
% Missed	16.7	0		
Reduction Analysis				
Total no of symbols	7735	1365	24 (else, 48)	4 (else, 8)
Time taken to run	7 mins	<= 7 mins		
No of terminals	340	<= 340	< 340	< 340
% True positives	60.7	62.5		
% False positives	39.3	37.5		
% Damaged	0	0		
% Detected	66.7	100		
% Missed	33.3	0		

Fig. 5.75 Analysis of Fig. 5.13

	Multiple Thresholds	*Empirical Pruning*	*Symbolic Pruning (for 1 vocabulary symbol)*	*Symbolic and Empirical Pruning (for 1 vocabulary symbol)*
Total no of dimension_descriptions: 4				
Complete Analysis				
Total no of symbols	265371	37543	30 (else, 60)	4 (else, 8)
Time taken to run	1hr 19mins	<= 1hr 19mins		
No of terminals	1086	<= 1086	< 1086	< 1086
% True positives	45.5	48.6		
% False positives	34.8	40.3		
% Damaged	19.7	11.1		
% Detected	75	75		
% Missed	25	25		
Reduction Analysis				
Total no of symbols	49557	10522	18(else, 36)	2 (else, 4)
Time taken to run	1hr 6mins	<= 1hr 6mins		
No of terminals	813	<= 813	< 813	< 813
% True positives	45.8	47.2		
% False positives	26.7	41.6		
% Damaged	27.5	11.1		
% Detected	75	75		
% Missed	25	25		

Fig. 5.76 Analysis of Fig. 5.77

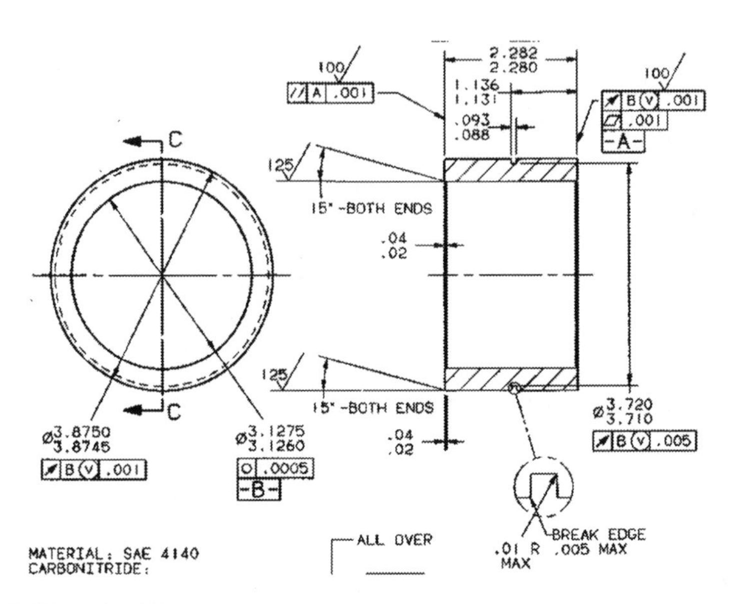

Fig. 5.77 Dimension 1(a)

As mentioned in Sect. 5.5.5.3, we prefer lower false negatives over lower false positives. Hence, even if at any time the false positives of reduction analysis are lower than the false positives of complete analysis, it is wise to choose the complete analysis for the above reasons.

Even if complete analysis produces better true positives, it still has poor performance with respect to reduction analysis in the *time* factor. The first method described in Sect. 5.5.5.2, could still be used but the second method is error prone. Hence, we need to find other better ways to decrease the time taken by the complete analysis.

The other observations made from Figs. 5.75 and 5.76 are as follows:

1. The performance shown in Figs. 5.75 and 5.76 were measured using six threshold pairs for *ANGLE_MAX* and *TOUCH_LEN*. It can be seen that there isn't much difference in run-time between the complete and reduction analysis using multiple thresholds, unlike the run time difference in the image analysis stage as shown in Table 5.9. This emphasizes the need for powerful output reduction strategies in the image analysis stage.
2. In the analysis with empirical pruning, we applied the pruning only to the highest level vocabulary symbol *dimension_description*. Hence, the run time is close to the run time without empirical pruning. Had the pruning been applied to the lower level vocabulary symbols, out of the six different results (from six different threshold pairs), five could have been ruled out much earlier which could have decreased the run time considerably.
3. We haven't measured the run-time using symbolic pruning (with/without empirical pruning) as we developed the 0-form grammar of G only for a specific

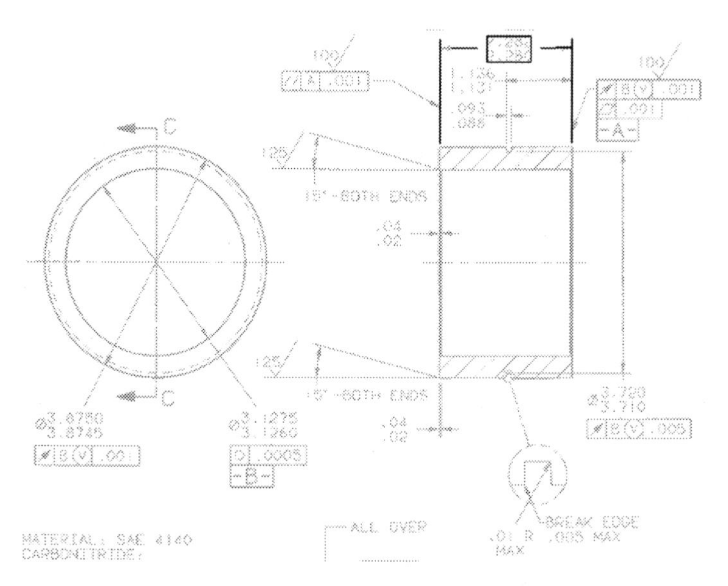

Fig. 5.78 Drawing 1

rule (the rule that produces *symmetric_pointer_pair_out*) and not for others. As shown, the number of symbols detected is reduced to half for this specific rule. Many of the rules that interest us are symmetric in nature, and, thus it can be safely presumed that the run time would decrease by a considerable amount.

4. If empirical pruning were coupled with symbolic pruning, the run time could be decreased to as low as one-sixth of the complete analysis run time.

From the above observations, it can be concluded that complete analysis is much preferred over reduction analysis. Within complete analysis, symbolic pruning combined with empirical pruning would yield better performance in both increasing the percent detected as well as decreasing the run-time of structural analysis.

5.5.7.2 More Analysis

The NDAS analysis was run on four more images and their results are presented in this section. Since it was argued in the previous section that the complete analysis is preferable to reduction analysis, the following analysis has not been done for the reduction case (Figs. 5.78–5.98).

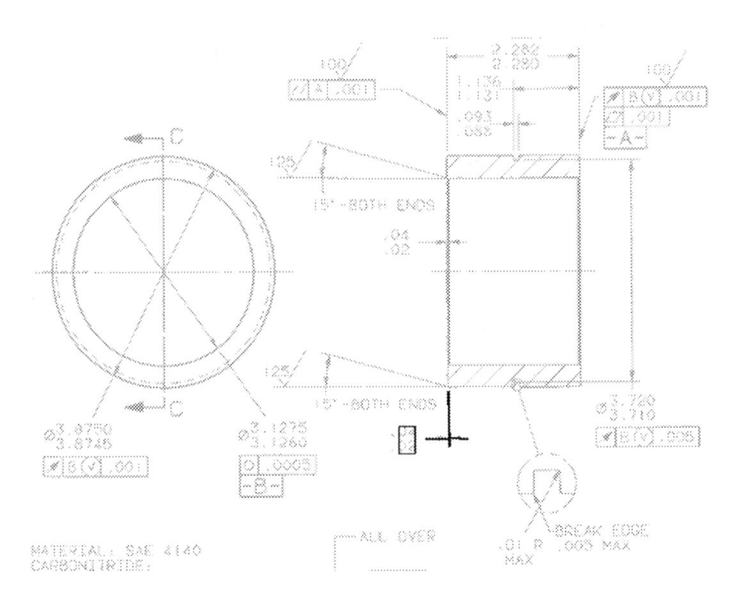

Fig. 5.79 Dimension 1(b)

	Multiple Thresholds	*Empirical Pruning*	*Symbolic Pruning (for 1 vocabulary symbol)*	*Symbolic and Empirical Pruning (for 1 vocabulary symbol)*
Total no of dimension_descriptions: 12				
Complete Analysis				
Total no of symbols	27192	7005	200 (else, 400)	50 (else, 100)
Time taken to run	2hr 20mins	<= 2 hr 20mins		
No of terminals	1492	< 1492		
% True positives	49.5	55		
% False positives	50.5	45		
% Damaged	0	0		
% Detected	16.6	16.6		
% Missed	83.4	83.4		

Fig. 5.80 Analysis of drawing 1

Fig. 5.81 Drawing 2

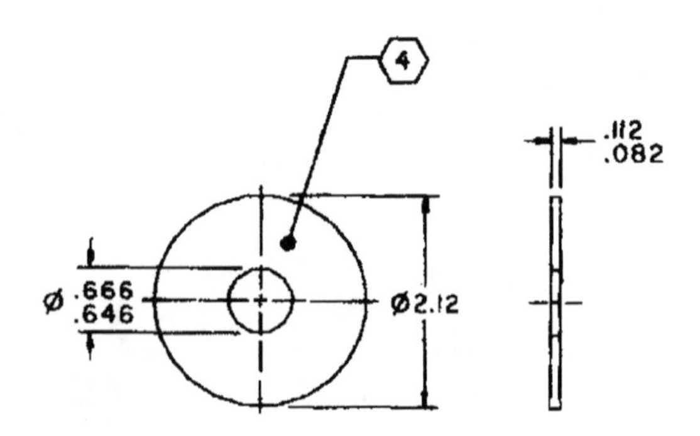

Fig. 5.82 Dimension 2(a)

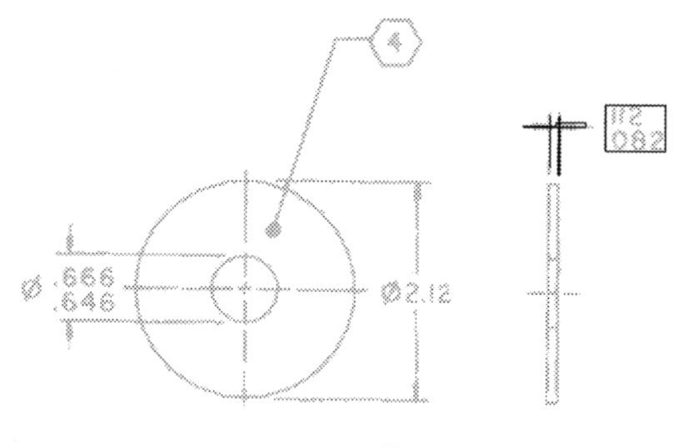

Fig. 5.83 Dimension 2(b)

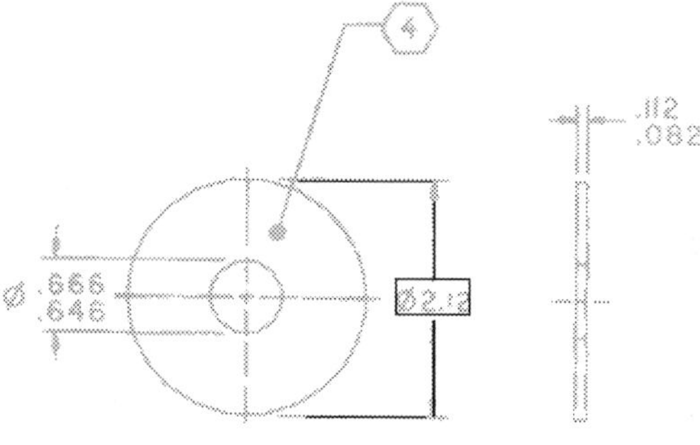

Fig. 5.84 Dimension 2(c)

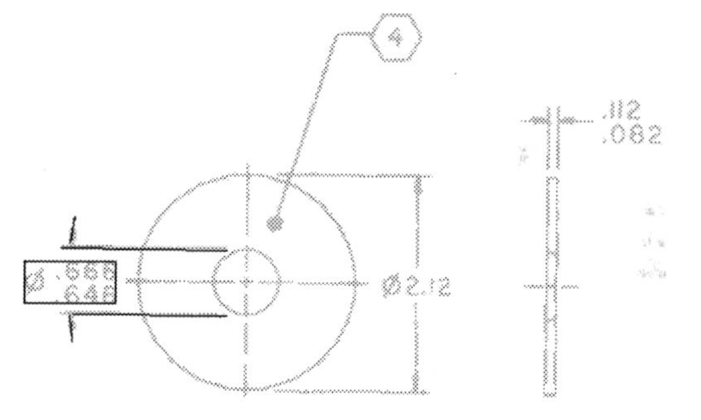

Most of the *dimension_descriptions* went unrecognized since the rules to detect them were not setup. The *dimension_descriptions* involving the **pointerarc_lines** were not detected as the **pointerarc_line** went undetected. Also, one of the **pointer_lines** embedded in the **circle** is slightly disconnected. Thus, it also was not detected by the arrow detection algorithm.

All the *dimension_descriptions* could be detected but the analysis also gave rise to 50 % of damaged results. This is because, in many of the results, the text wasn't detected wholly.

The above drawing took the least analysis time of the whole lot since it is relatively noise free and smoothly connected. Nevertheless, many of the *dimension_descriptions* could not be detected because, either the rules were not supportive or some of the **pointer_rays** forming the *dimension_description* were undetected

	Multiple Thresholds	*Empirical Pruning*	*Symbolic Pruning (for 1 vocabulary symbol)*	*Symbolic and Empirical Pruning (for 1 vocabulary symbol)*
Total no of dimension_descriptions: 3				
Complete Analysis				
Total no of symbols	171692	43350	88 (else, 176)	22 (else, 44)
Time taken to run	1hr 40mins	<= 1 hr 40mins		
No of terminals	7854	< 7854		
% True positives	76	81		
% False positives	26	19		
% Damaged	50	50		
% Detected	83	100		
% Missed	17	0		

Fig. 5.85 Analysis of drawing 2

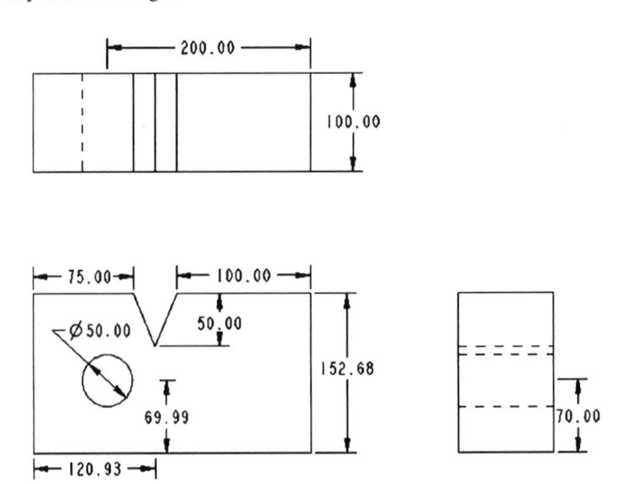

Fig. 5.86 Drawing 3

by the image analysis agents. The reason for the former case was, many of the *dimensions* in the image refer to objects in another view of the image. But the rules present so far, don't detect such relations.

Two of the *dimension_descriptions* were not detected as they contain very small arcs in them, and the rules written so far do not support arcs as terminals. The *dimension_description* containing the datum_reference was detected with partial text (hence, damaged).

5.6 Explicit and Persistent Knowledge

Domain knowledge permeates all aspects of the analysis, including the physical processes operating on the paper drawing, e.g., printing, folding, staining, etc., the image analysis, e.g., the notions of points, lines, blobs, and the particularities

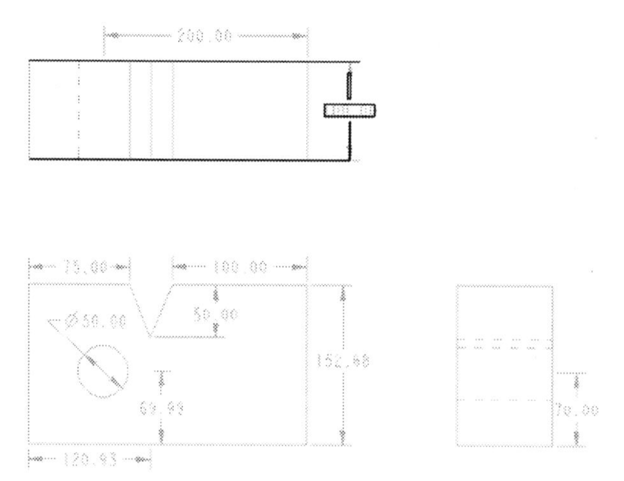

Fig. 5.87 Dimension 3(a)

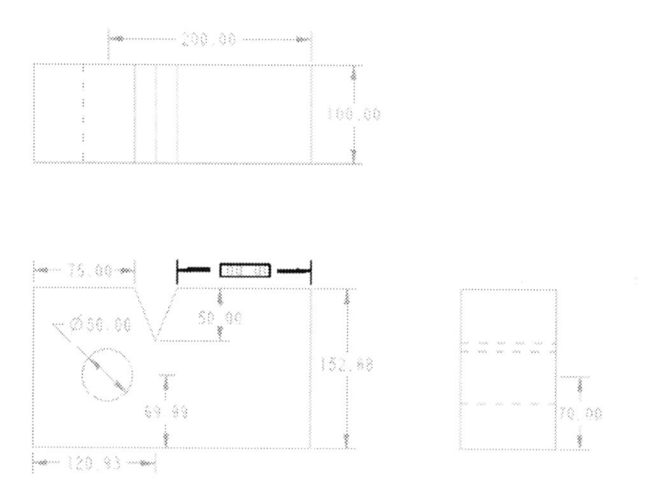

Fig. 5.88 Dimension 3(b)

of the algorithms that are implemented for their recovery, the structural layout and interpretation semantics of the contents of the drawing, and eventually, the broader context within which the drawing analysis will be exploited in the reverse engineering setting. Here we discuss the nature of such knowledge, how it can be made explicit (both for agents and humans), and how performance models can be defined, calibrated, monitored and improved in terms of this knowledge. A framework is proposed that allows the user or agents to: (1) explore the threshold space for an optimal drawing analysis, (2) control acquisition of new data (e.g., view token generation as state estimation and select agent actions that optimize information gain), (3) incorporate knowledge in abstract form and communicate

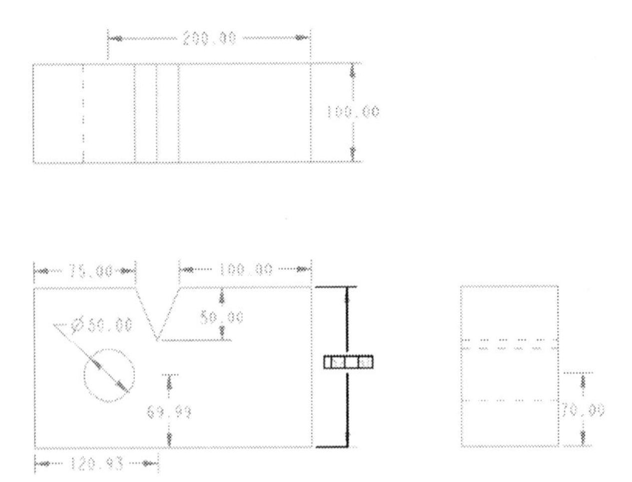

Fig. 5.89 Dimension 3(c)

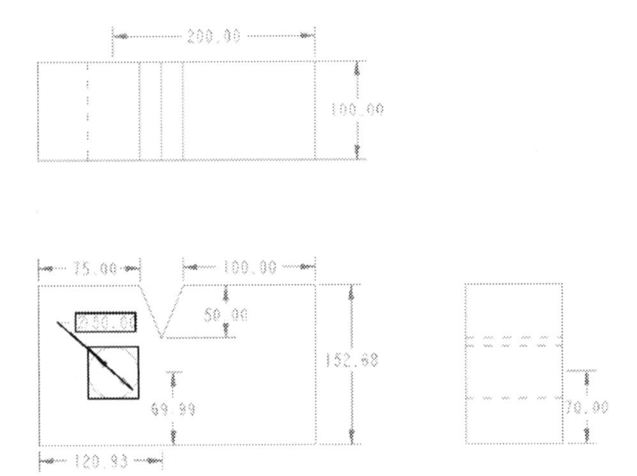

Fig. 5.90 Dimension 3(d)

abstractions between agents and users, and (4) inform software engineering and system development with deep knowledge of the relationships between modules and their parameters (at least in a statistical sense).

The reverse engineering of legacy systems is a difficult and complex problem, but vital in certain domains. This usually involves a physical instance of the system, as well as some paper drawings produced by hand or from mechanical CAD systems. The goal may range from producing a replica, to changing some parameters, to a major re-design. For example, Fig. 5.99 shows a gearbox that operated for many years onboard a ship, and then failed. Developing reverse engineering techniques from such a physical example and any available related engineering drawings is our goal.

	Multiple Thresholds	*Empirical Pruning*	*Symbolic Pruning (for 1 vocabulary symbol)*	*Symbolic and Empirical Pruning (for 1 vocabulary symbol)*
Total no of dimension_descriptions: 10				
Complete Analysis				
Total no of symbols	110638	28333	372 (else, 744)	93 (else, 186)
Time taken to run	1 hr	< 1 hr		
No of terminals	2399	< 2399		
% True positives	54.4	63		
% False positives	45.6	37		
% Damaged	0	0		
% Detected	35	40		
% Missed	65	60		

Fig. 5.91 Analysis of drawing 3

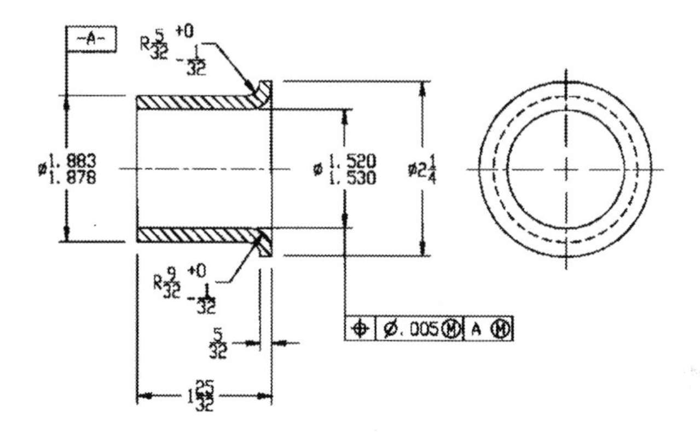

Fig. 5.92 Drawing 4

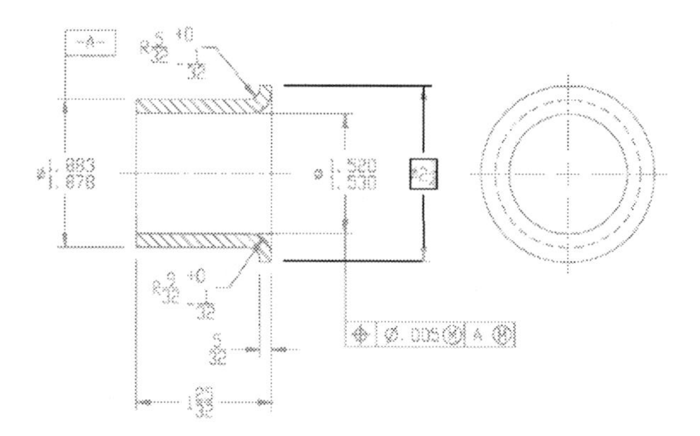

Fig. 5.93 Dimension 4(a)

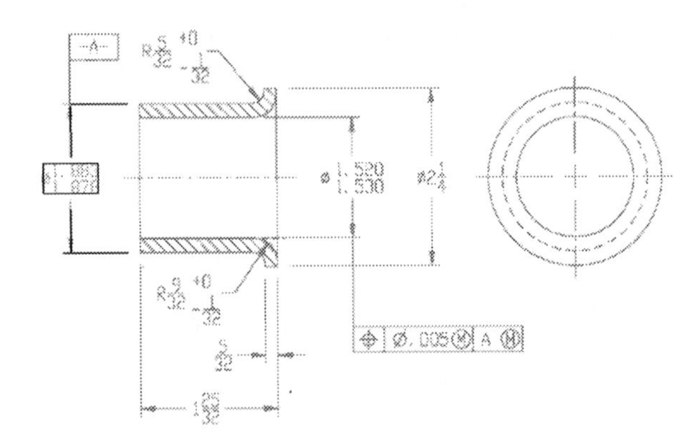

Fig. 5.94 Dimension 4(b)

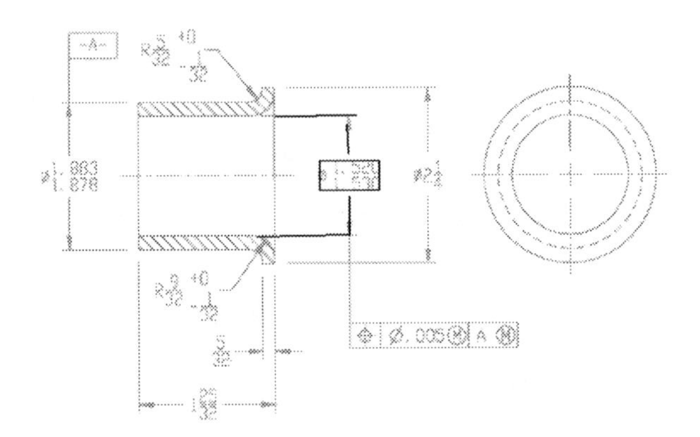

Fig. 5.95 Dimension 4(c)

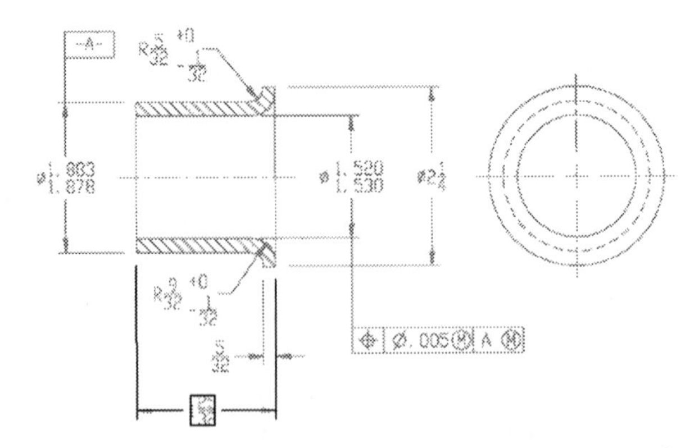

Fig. 5.96 Dimension 4(d)

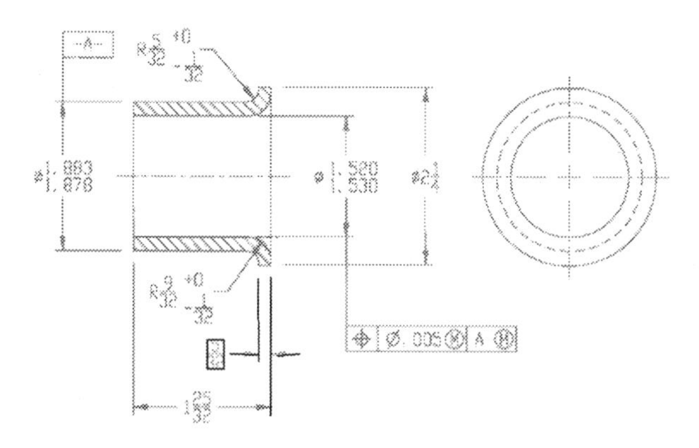

Fig. 5.97 Dimension 4(e)

	Multiple Thresholds	*Empirical Pruning*	*Symbolic Pruning (for 1 vocabulary symbol)*	*Symbolic and Empirical Pruning (for 1 vocabulary symbol)*
Total no of dimension_descriptions: 7				
Complete Analysis				
Total no of symbols	34444	5231	200 (else, 400)	50 (else, 100)
Time taken to run	1hr 20mins	<= 1 hr 20mins		
No of terminals	1307	< 1307		
% True positives	66.5	70		
% False positives	33.5	30		
% Damaged	24	43		
% Detected	76	57		
% Missed	28.5	28.5		

Fig. 5.98 Analysis of drawing 4

Fig. 5.99 Newport news gearbox to be reverse engineered

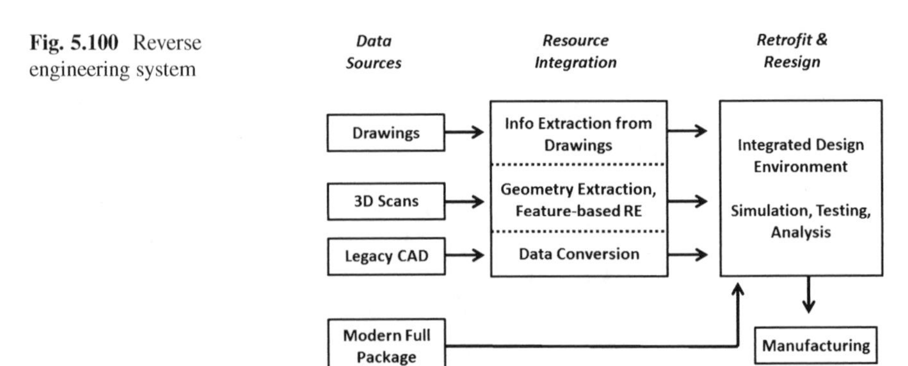

Fig. 5.100 Reverse engineering system

Fig. 5.101 Envisioned virtual interface to model surface, point cloud and drawing data

Figure 5.100 shows the overall reverse engineering system we are developing; the goal is to take advantage of data about the system in all its forms: drawings, 3D scans, and CAD models as they are constructed, as well, and to allow the user virtual access during the redesign process (see Fig. 5.101). The wider knowledge involved includes manufacturing information and constraints, design analysis codes (e.g., stress or aerodynamics), cost/performance models, etc.

5.6.1 *Engineering Drawing Analysis with NDAS*

We have shown that a structural model may be realized through a set of software agents acting independently and in parallel to ultimately achieve a coherent analysis of CAD drawings [44, 48, 49, 113]. The high-level goals of the analysis are to:

- Understand legacy drawings.
- Acquire context of field and engineering data.
- Respond to external analysis, user input.
- Integrate drawing analysis in redesign.

NDAS allows multiple agents to produce the same type of data, for example, line segments or text. Other agents which use these entities as inputs may choose from any or all of the available sets of data to produce their own data. Moreover, even a single agent can produce its output using multiple thresholds, or can be asked by another agent to produce output with a given set of control parameters. This allows people or more sophisticated agents to explore the entire parameter space of all the agents involved in the analysis.

The mechanism to handle the combinatorial explosion of data is tied to the structural definition of the engineering drawing, and uses syntactic analysis to eliminate redundant comparisons. This symbolic redundancy calculation uses both the syntax of structural re-write rules, as well as parsing constraints on the tokens generated from the image analysis to achieve orders of magnitude reductions in the possible combinations of tokens. However, NDAS to date has done little else to incorporate or exploit the wealth of other knowledge involved in understanding engineering drawings.

5.6.2 *Knowledge About Engineering Drawing Analysis*

Figure 5.102 shows the sequence of paper drawing creation and exploitation with which we are concerned. We consider knowledge about physical processes, image analysis and document interpretation.

5.6.2.1 Physical Processes

It is important to capture knowledge about all aspects of the physical processes involved. For example, printing gives rise to certain errors that can influence the image analysis and subsequent interpretation. During storage and usage, it is possible to introduce lines by folding or creasing, or to obscure lines and text by stains, writing or damage to the paper. Scanning is itself a physical process subject to motion blur, lighting, scale and other perturbations. Good understanding

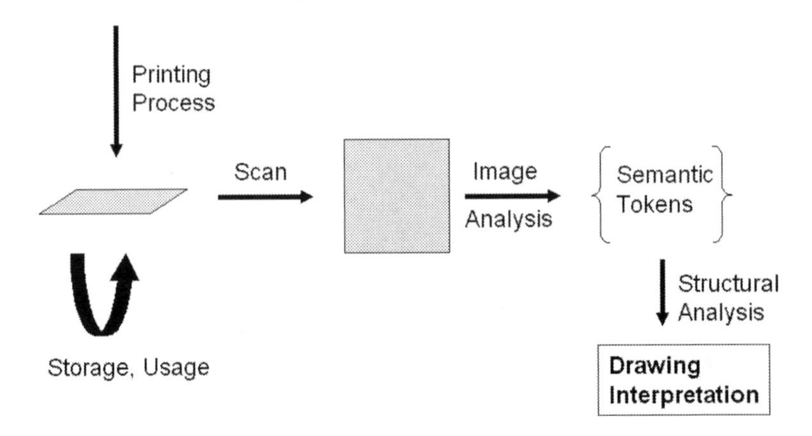

Fig. 5.102 Engineering drawing analysis process

is necessary for robust and correct analysis, and a good synthesis model will allow the controlled creation of test data with defects.

5.6.2.2 Image Analysis

Discrete geometry plays a large role in the analysis of engineering drawings, and involves abstract notions, including:

- 0-dimensional objects: isolated points, corners, branch points, end points, etc. and relations: distance, near, same kind, etc.
- 1-dimensional objects: line segments, straight segments, circles, boxes, etc. and relations: collinear, parallel, perpendicular, neighbor, closed, etc.
- 2-dimensional objects: blobs (e.g., arrowheads) and relations: above, left of, touches, occludes, etc.

Moreover, these notions cannot be implemented perfectly, and it is important to know how the realizations differ from the ideal (e.g., what's the threshold for parallel?). Even more important is the relation of these notions and their recovered approximations to the semantic tokens which form the basis for the structural analysis.

5.6.2.3 Structural Analysis

The structure of the drawing is given by a set of tokens (e.g., line segments, text, pointers, graphics, manufacturing symbols, etc.) and the relations that hold between them. Thus, the production of the tokens is crucial, and interpretation problems arise when tokens are missing, broken into parts, or falsely reported. The relations between the tokens need to be clearly defined, as well as the amount of divergence from the ideal. Context of various sorts is also extremely important, and ranges from geometric frame (which way is up?) to drawing type (detail drawing, assembly description, manufacturing constraint requirements, etc.).

These various sets of knowledge are usually not made explicit, either during the development of the system or for exploitation during an analysis. We are interested in answering the following kinds of questions:

1. How can this knowledge be made explicit?
2. How can the differences between the ideal and the implementations be given?
3. Can some of the knowledge (ideal or performance) be learned by the agents?
4. How can people interact with this knowledge to understand why the system does something or to change how the system does it?
5. How can the knowledge be exploited during the analysis of one image; over a set of related images? over various projects? i.e., in order to gain and record more insight on engineering drawing analysis in the long term.

It is essential to answer these questions so that the system can improve over time, and be more effectively understood and exploited by its human operators.

5.7 Proposed Method

We propose the following approach to address this problem:

1. Give a specification for the ideal.
2. Give ways that implementation can differ from ideal.
3. Give a measure of the difference.
4. For every analysis, keep a record of the ideal referent, actual produced, difference measure and analysis parameters.

For example, parallel segments should ideally have 0 degrees difference in angle. A difference measure would be the actual difference in angle, or some monotonically increasing function (square, exponential, etc.). Various implementations would carry different information; e.g., if parallel is computed from the two segment angles, then an angle difference threshold would be kept; if parallel is determined by whether the points defining the one segment are all the same distance from the other segment, then the maximum and minimum distances would be kept. It is possible to have agents for both parallel operators, and the system can decide (based on training or operator feedback) which is better. This goes with our notion to develop a system which allows many different analysis methods in parallel, and from this wealth of data, chooses between them to construct the best interpretation possible.

This approach also fits well with statistical methods. For example, various information measures can be defined and used to steer the analysis. Once we have established mechanisms for knowledge expression and use, we will explore alternative mechanisms for the exploitation of that knowledge (for example, Durrant-Whyte and colleagues [39] have developed methods to maximize information gain with each observation action—this approach might give good results here).

5.7.1 Knowledge About Engineering Drawings

Let's look in more detail at the knowledge that would be useful in this application. As for engineering drawings per se, the useful information concerns:

- *Layout*: which way is up? down? Usually represented as semantic network or a grammar.
- *Symbols*: alphabets, digits, special cases. Can be expressed with dictionary, images, or networks.
- *References*: includes conventions for pointers, names, use of circles, etc., and can be described with semantics net or as image features.
- *Characters*: language, numbers, measures; implemented as semantic net, feature vectors or images.
- *Real world semantics*: manufacturing information, 3D, 2D projections, etc.; typically given as semantic net.

As can be seen, most of this knowledge, if it exists, might be better expressed as a semantic network or in vector or image form. We are currently investigating the construction of a domain ontology , and hope to base it on the Standard Upper Merged Ontology (or SUMO) [90]. In this way, we make the assumptions of the agents explicit, and provide a SUO-KIF [36] interface to other users and systems. However, it must be pointed out that our domain requires analogical forms of knowledge as well, including: images, 3D data sets from Coordinate Measurement Machines or laser scanners, etc. Some axiomatizations and ontologies for geometry exist (e.g., see [11, 99, 117], but their usefulness in this context remains to be seen.

Image analysis has its own set of concerns, including:

- 1D segments,
- pixels (digitization),
- relations, and
- realization of geometry.

Algorithms include: thresholding foreground/background, thinning, segment extraction, straight segment determination, geometric objects detection (e.g., boxes, circles), pointer detection, and text detection. Each of these must deal with thresholds, sensitivity analysis, quality estimates, complexity, and robustness with respect to other algorithms.

Finally, knowledge about goals may influence agent actions; here are some goals that the system may be asked to achieve:

- Find part name.
- Find label information.

- Extract references to other parts.
- Get dimension information for specific part features.
- Determine manufacturing constraints.
- Determine safety or other special descriptions in the text.

These various forms of knowledge should not be static, but should be adjustable over time, as more experience is gained. For example, the use of pointers in drawings can be quite creative, and these need to be cataloged and accounted for. At a minimum, threshold exploration should be possible and recorded.

Another issue is what needs to be communicated between agents (and/or users) which includes at least the following:

- the goal,
- the results of an agent; this includes the info produced, info about the production of the info, and some quality of result measures, and
- feedback to an agent; for example, *this data resulted in no solution* or *parallel constraint needs to be tighter* or *your results are not necessary for this goal*; this last feedback would lead to greater efficiency if agents know when they are unnecessary.

For example, the circle agent uses simple 1D segments (a set of pixels) as input and checks if the set of pixels forms a circle. However, this agent is not necessary for the analysis of the title block of a drawing; it is essential, however, for full drawing analysis. The result of the analysis is a list of point sets determined to constitute circles, and for each circle gives the center and radius, the segments or pixels involved, a quality measure of the circle, and the resources used to produce the circle (e.g., data files used, space and time complexity, etc.). It may also be necessary to include information about why the thresholds and parameters were selected. As an example of feedback that the circle agent may want to provide, suppose that it uses straight line segments to detect circles (i.e., a set of straight line segments form a circle if they are connected end to end and their points do not lie too far from a circle); if the straight segments are fit too coarsely, they may not form a circle, when in fact the pixel data would permit a circle. Thus, the circle agent may want to ask the segment agent to re-fit the data with a tighter linear fit threshold.

As a starting point, we have investigated the knowledge about thresholds and their interplay between entities produced, consumed, and the semantic tokens generated. Figure 5.103 shows the image analysis part of NDAS. Threshold utilization is indicated by the circled numbers. The meanings of the thresholds can be given as follows:

- **Circle 1**:

We term the image analysis knowledge given in Fig. 5.103 as superficial knowledge, since it concerns only the external relations between the agents and their products. Thus, information about the organization of modules, which use the data from which others, their production information, the quality measures on the data, the amount and trends of data production, and the system activity all fall under this term (Table 5.10).

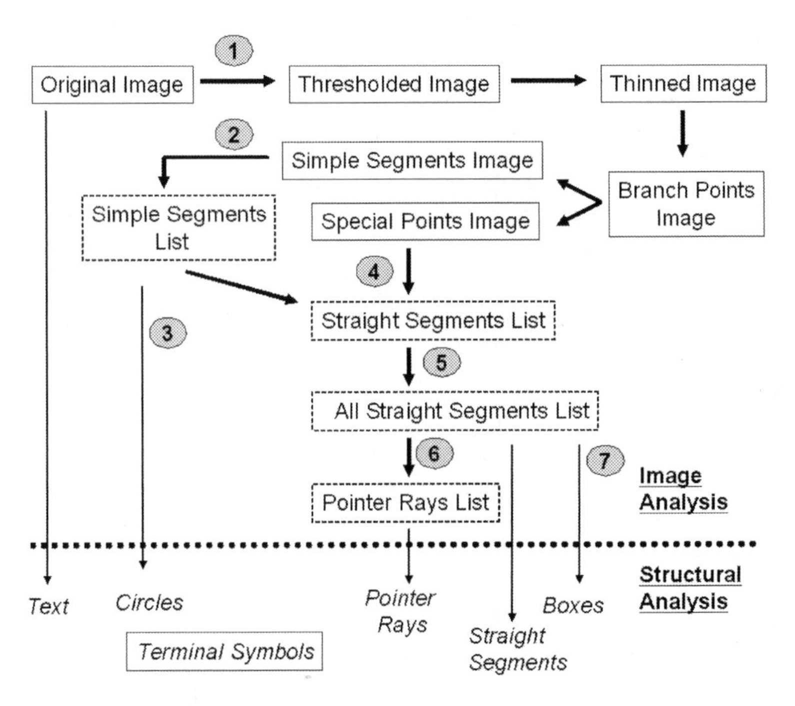

Fig. 5.103 The image analysis agents and the flow of data between them

Table 5.10 Image analysis agent thresholds and parameters and their impact on Fig. 5.103

Circle no.	Thresholds/parameters
1	Foreground/background
2	Pixel curvature parameters
3	Circle fit parameters
4	Line fit parameters
5	Collinear; line fit parameters
6	Endpoint distances; segment lengths; collinear
7	Segment length, separation threshold, parallel, perpendicular duplicate threshold

Circle no.	Related impact
1	Extra/missing pixels; connectivity of segments
2	Corner detection, straight segment endpoints
3	Circle detection, reference detection
4	Number and quality of segments
5	Large-scale object detection
6	Pointer ray detection, dimension analysis, references
7	Box detection; document block analysis; text analysis

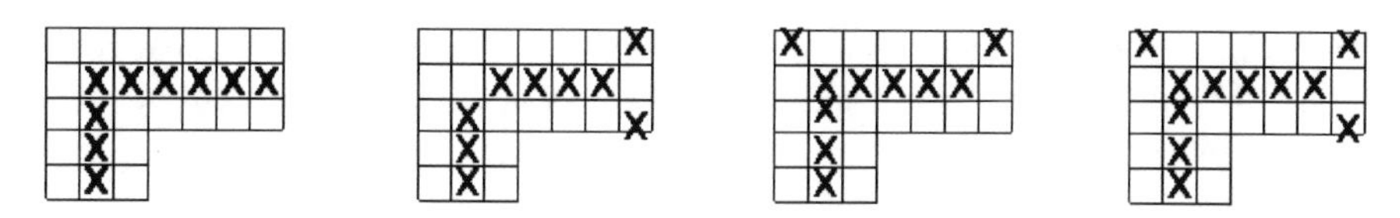

Fig. 5.104 Four variations of a thinning operation

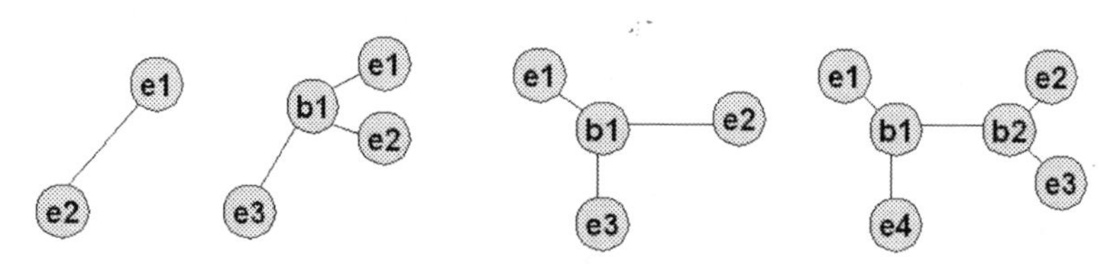

Fig. 5.105 Graphs of connectivity between end points (e1,e2,...), and branch points (b1,b2,...) of thinned objects from Fig. 5.104. (Between every pair of nodes is a, not necessarily straight, line segment.)

Opposed to that is deep knowledge, which concerns the inner workings and decision rationales for implementations, threshold settings, etc. This then includes an ideal description of the process, an explanation set of how the implementation differs from the ideal, a characterization of the likelihood of the variances from the ideal, and the relation of the variations to further processing, including semantic token (terminal symbol) creation and semantic analysis.

To clarify these ideas, consider the image thinning process. There is a mathematical notion of a valid thinning operator on point sets, but implementations may vary from this ideal for many reasons and with different implications. Consider the four versions of the thinned partial segment in Fig. 5.104.Which of these is produced may significantly impact later analysis; e.g., abstractions based on end point, branch point and straight line segment relations can be radically different. Figure 5.105 shows a set of relation graphs for the thinned objects above. As can be seen, the number of line segments, the position of their endpoints and the geometric relations between them (distance, parallel, etc.) can all be greatly affected by these differences in the thinning. Thus, what might be viewed as a local or minor algorithm issue, may lead to a radical change in performance (including increase in complexity if lots of small segments are generated) if there is no knowledge of how one process impacts other processes through shared analysis objects. It is of great interest to understand these relationships, and to declare them when the system is designed and implemented, but even if that is not possible or accurate (the developers may not understand the impact!), it would be good to allow the system to determine some of this knowledge as various algorithms are executed with different parameter values.

In terms of the thinning operation, we might proceed as follows:

Ideal definition of thinning One example of this is the medial axis transform [15]. This is the set of points such that a circle centered at the point touches the boundary of the object in at least two distinct places.

Algorithm difference from ideal the algorithm may approximate the ideal definition in order to reduce computational complexity and because the ideal notions don't apply perfectly to digital geometry. The following differences may occur:

1. Ends of segments may be fragmented.
2. Corner regions of segment may be fragmented.
3. Medial axis may be displaced from actual corner location.

Measures of difference Several possibilities exist to measure the three differences listed above. There are two levels of measure, however. First, it is of interest to measure individual errors in terms of the number of extra segments produced, or the distance a thinned set is displaced from a point of interest in the original point set. In addition, it is useful to have some statistics over the whole population. For example, this might be either (1) a likelihood on the number of extra fragments expressed as a mean and variance or in other forms, or as a function of the original segments, the features of the segment or those of the thinned segment. For example, if the thinned segment is perfectly straight, then it is most likely that it perfectly represents the ideal.

Model Calibration This approach also affords the opportunity to generate controlled test data and to obtain very good estimates of how well the model works. This would work as follows. A CAD model is developed for some artifact. This is then printed, possibly submitted to various degradations, and then scanned. Since the actual CAD model is available, it is possible to know the perfect set of pixels that should have been printed and then scanned. Once the thinned objects are determined, they can be compared to the perfect set of thinned objects, etc. We call this *model calibration*, as it can be used to determine how well the process model measures the true state of affairs.

5.8 Examples

We have performed many experiments with the image analysis part of NDAS. Figure 5.106 shows part of a typical scanned engineering drawing, and the thinned image in Fig. 5.107. One thing to notice is how the arrowheads in the original image have been changed into line segments. Also, the corners of boxes have been displaced several pixels from where they should ideally be located. Figure 5.108 shows the boxes detected in the image, so it can be seen that it is still possible to find them, however, this may cost a great deal in computational or algorithmic complexity, or the algorithms may in fact be tuned for one image and not work very well on another. This is the kind of knowledge we would like to gain and record for better exploitation of the system. A change in thresholds of two pixels in length, and parallel segment overlap of 10 % more, results in missed boxes.

Now consider in detail the kind of information to be gathered and characterized about the thinning algorithm. (Note: the ground truth locations of corner points for

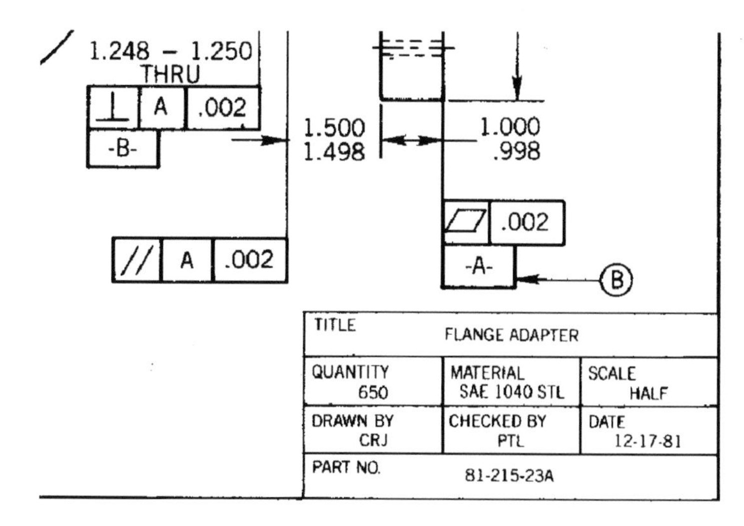

Fig. 5.106 Part of a typical scanned engineering drawing

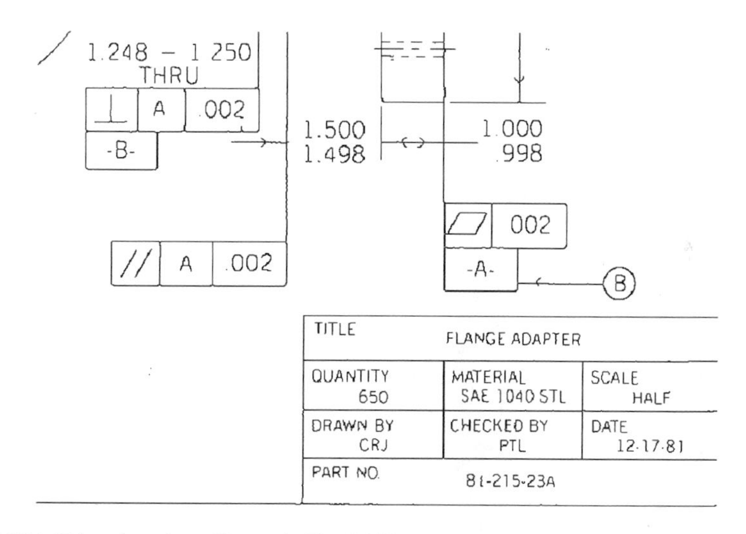

Fig. 5.107 Thinned version of image in Fig. 5.106

the boxes in the image have been given by hand.) We would like to model the impact of the algorithm on:

1. True corner existence.
2. Segment recovery (particularly, endpoint location).
3. Box detection and localization.

For the image in Fig. 5.106, the histogram in Fig. 5.109 shows the ideal corner points distance from the thinned pixel set. The segment endpoint distance histogram is given in Fig. 5.110. This data is for ideal segments such that there exists a segment

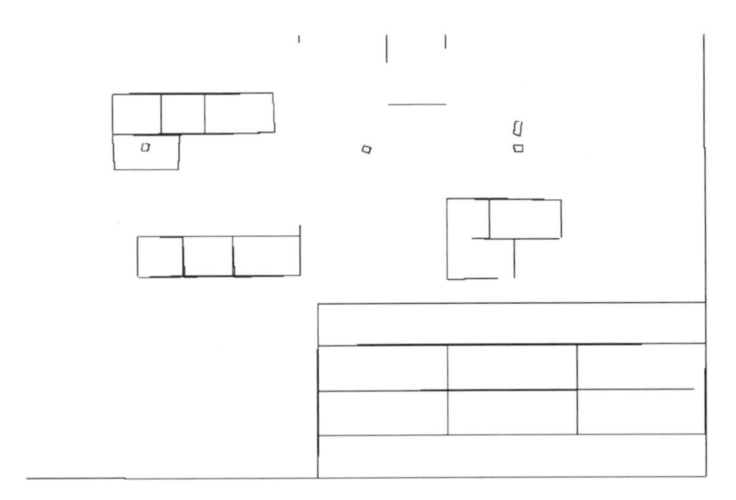

Fig. 5.108 Boxes found in thinned image

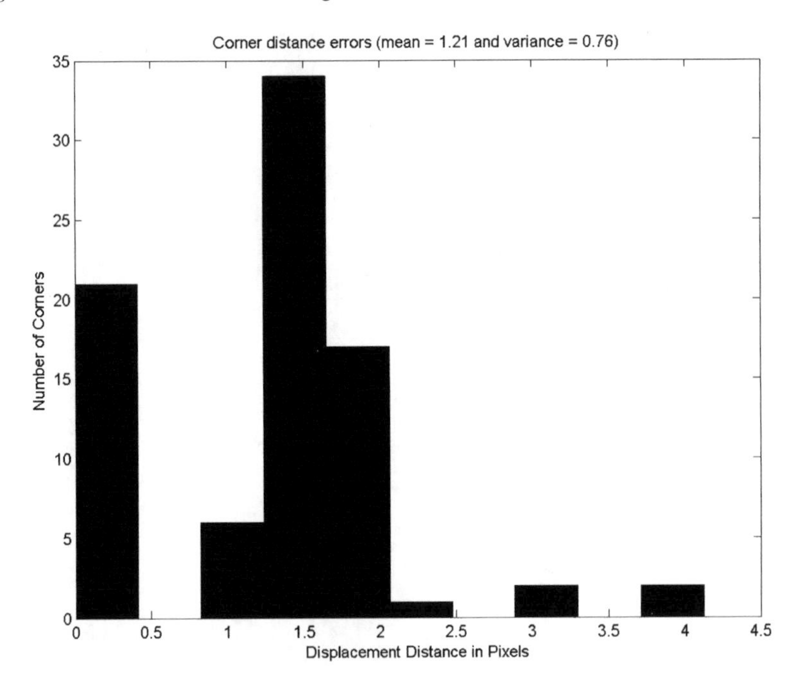

Fig. 5.109 Histogram of ideal corner point displacement in thinned image

produced by the image analysis whose endpoints are within ten pixels of the ideal segment. (The number of missing segments is eight; i.e., eight ideal segments have no counterpart in the segments extracted from the image.) Figure 5.111 gives the histogram for the distance of ideal box point corners from detected data.

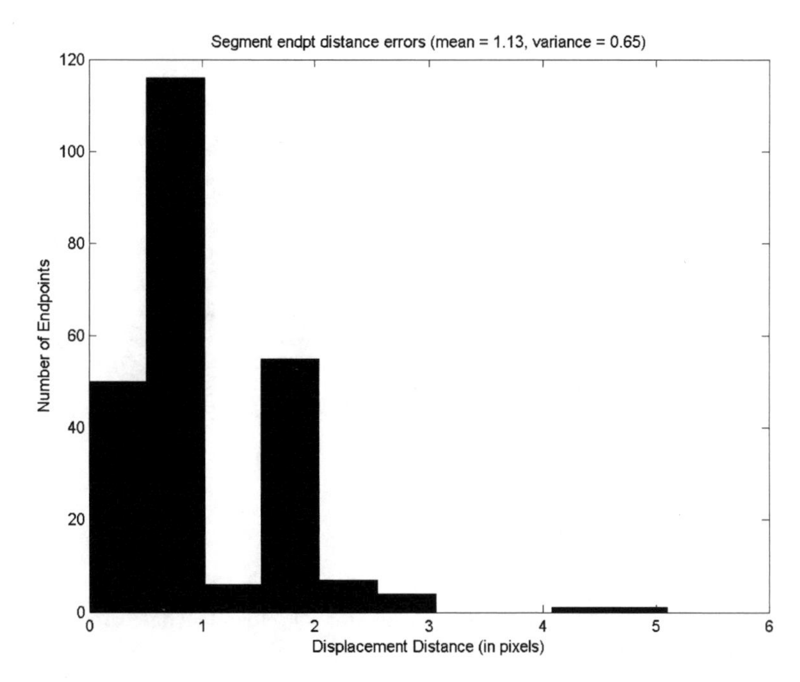

Fig. 5.110 Histogram of ideal box segment endpoint distances from detected

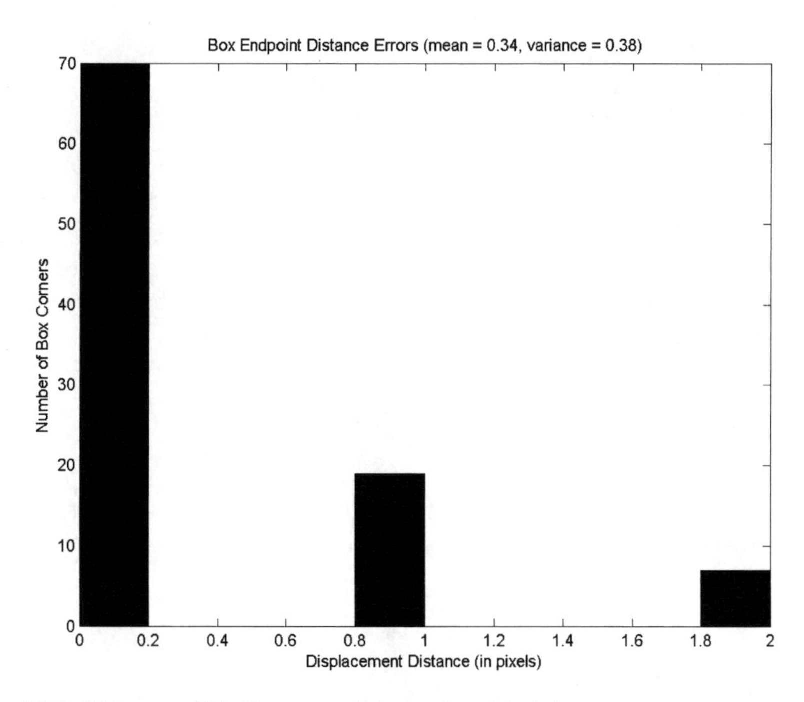

Fig. 5.111 Histogram of ideal box corner distances from detected

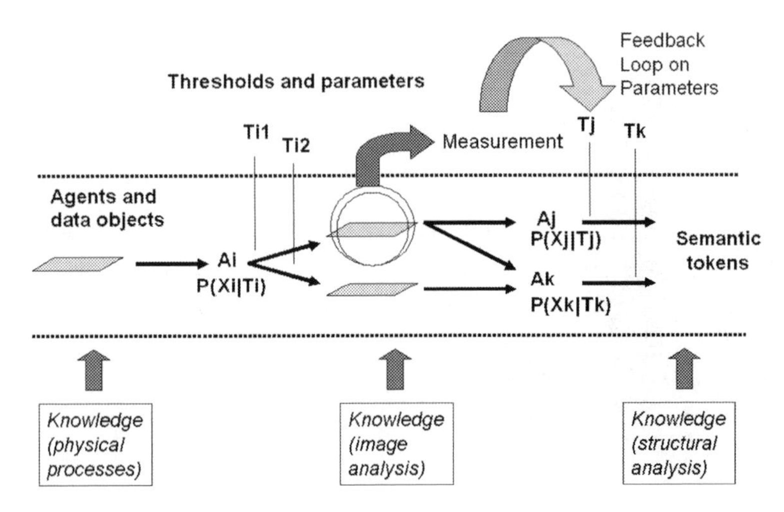

Fig. 5.112 Smart agents network system

All the ideal boxes were found, and the error is very low. The data shows that the box detector algorithm is insensitive to endpoint displacement in the thinning and segment detection algorithms. Moreover, even a missing segment does not preclude the detection of a box, so long as there is a reasonably long segment found on each side. This depends, of course, on the thresholds in the box detector agent. (Also, note that the box agent discovers *boxes* in the image that are not included in the ideal set; e.g., the upper part of a letter "B" in the text.)

We have to this point tried to convey a sense of the kinds of knowledge that interest us, and how they can be used in engineering drawing analysis. We now give a high-level summary of our proposed theoretical framework and enumerate some advantages that may result from this approach.

Figure 5.112 shows a set of agents, A_i, each of which produces various outputs using a set of parameters and thresholds, T_i, and each having an associated model (or set of models), $P_i(X_i|T_i)$, describing the agent's variance from the ideal in terms of some appropriate measure. Knowledge of three sorts (physical, image analysis, and structural interpretation) is available and informs the agents' actions and understanding of each others results. Higher-level control processes may exploit this in several ways:

1. Explore the threshold space for global optima (see feedback loop in Fig. 5.112).
2. Control acquisition of new data (e.g., view token generation as state estimation and select agent action that optimizes information gain).

3. Incorporate knowledge in abstract form and communicate abstractions between agents and users.
4. Inform the software engineering and system development with deep knowledge of the relationships between modules and their parameters (at least in a statistical sense).

The current status of the project (called the Smart Agent Network System or SANS) is that the core image and structural analysis components have been developed and applied to engineering drawing analysis to gain experience and insight into crucial agents and parameters. We are now exploring the representation of this domain knowledge in specific nomenclatures. We are also investigating state estimation frameworks to provide a more incremental analysis based on observations provided by the system, and the associated information measures (see [18] for an introduction to the area). Notice that each program execution can be viewed itself as a measurement on the image, and the set of measurements will be used by a control process to achieve the best interpretation of the drawing.

A larger issue is the use of other types of information in the reverse engineering scenario; e.g., 3D scanner data, photos, manufacturing information, etc. These analogical forms of data must be integrated into the re-design as well, and this should be done so as to allow rapid iteration, and fast exploration of the design space.

Chapter 6
Map Background and Form Separation

Our goal is the extraction of semantic content from raster map images which may then be exploited in a variety of ways; e.g., the analysis of aerial imagery. The major map features that we want to extract include: roads and road intersections, texture, text, and miscellaneous symbols. Where appropriate these may be vectorized, and the image coordinates may be recorded as well. Meta-data may be provided including the type of map (e.g., USGS), any specific color encoding of features (e.g., USGS maps may use from 6 to 13 specific colors), geo-spatial information, and names of features may be known and sought in the image (e.g., rivers, lakes, roads, towns, etc.).

The specific goals in the next few chapters are to describe methods to segment:

- *Background*
  - Use of knowledge for color use in known map types.
  - Use of color probes: interactive and neighborhood analysis.
  - Machine learning methods trained on scanned maps and other sources.
- *Roads and Road Intersections*
  - Road extraction.
  - Road type determination.
  - Vectorization.
  - Sub-pixel road tracking.
  - Double line roads.
  - Road name and/or number.
  - Intersection recognition.
  - Geo-reference for intersections.
- *Texture*
  - Knowledge-base for constraints on texture appearance based on map type.
  - Long-term analysis to discover map textures and populate databases (e.g., $2D$ statistics, etc.).

T.C. Henderson, *Analysis of Engineering Drawings and Raster Map Images*,
DOI 10.1007/978-1-4419-8167-7_6, © Springer Science+Business Media New York 2014

- *Miscellaneous Features*
  - Highway markers.
    - Interstate markers.
    - US Highway markers.
    - State Highway markers.
  - Other (e.g., noise pixel sets).

6.1 Test Data

The algorithms can eventually apply to a wide range of map types: USGS DRG, NGA, and other miscellaneous types. The dataset that we use is described below. Tests are run with a variety of possible measures: (1) time cost per pixel, (2) true positive rate, (3) false positive rate, (4) false negative rate, and (5) the overall quality of the segmentation and interpretation.

6.1.1 USGS Maps

6.1.1.1 NGA Maps

The number of colors in a USGS map is limited to a few specific colors. These are described in Appendix C, and this information is exploited in the analysis of the USGA maps. Note that the number of colors used may differ from map to map (from 6 to 13 colors, including black).

6.2 Background Segmentation

The background of a map image consists of two major types:

- *Map Surround Area*: The areas surrounding the map proper, but not part of the map. This includes the legend, and border regions.
- *Map Embedded Background*: This is the underlying background of the actual map (typically **WHITE** pixels). This includes pixels that do not belong to any class found in the legend.

Directory	*File Name*	*Num Colors*	*Num Rows*	*Num Cols*
maps-early	madison.tif	9	5,750	6,951
maps-early	franklin.tif	9	4,672	5,725
Redstone Arsenal	O34086E5.TIF	11	5,563	6,916
Redstone Arsenal	O34086E6.TIF	9	5,563	6,909
Redstone Arsenal	O34086F5.TIF	9	5,571	6,912
Redstone Arsenal	O34086F6.TIF	9	5,572	6,910
Redstone Arsenal	O34086F7.TIF	9	5,555	6,897
Redstone Arsenal	O34086G7.TIF	11	5,571	6,919
Salt Lake City	q1120_drg24	12	4,369	5,731
Salt Lake City	q1121_drg24	12	4,361	5,726
Salt Lake City	q1122_drg24	248	8,705	11,438
Salt Lake City	q1219_drg24	12	4,386	5,738
Salt Lake City	q1220_drg24	12	4,377	5,731
Salt Lake City	q1318_drg24	12	4,402	5,744
Salt Lake City	q1319_drg24	12	4,394	5,738
Salt Lake City	q1320_drg24	12	4,385	5,738
Salt Lake City	SaltLakeCityDrgMap.tif	12	10,100	7,600
USGS DRG	F34086A1.TIF	5	10,599	6,003
USGS DRG	F34086E1.TIF	5	10,579	5,971
USGS DRG	mrg4619.tif	12	4,930	5,631
USGS DRG	nashville_east.tif	12	4,650	5,719
USGS DRG	wilsondrg.tif	12	22,844	20,268

Directory	*File Name*	*Num Colors*	*Num Rows*	*Num Cols*
NGA Maps	nimadata1018367581.tif	NA	1,682	1,486
NGA Maps	nimadata1018367594.tif	NA	3,363	2,965
NGA Maps	nimadata1021996034.tif	NA	5,605	5,940
NGA Maps	nimadata1024940339.tif	NA	1,122	1,486
NGA Maps	nimadata1024940385.tif	NA	2,242	2,965

6.2.1 Map Surround Background

Raster map images may or may not contain a *Map Surround Background*. Therefore, it is necessary to first determine if this exists. The main feature of the *Map Surround Background* is large areas of pure **WHITE** (index value equals 1 in the image; see Appendix C). If such areas are found, then the map proper must be segmented from the image for further analysis. In addition, the *Map Surround Background* can be analyzed for its semantic content. This includes the legend area and its contents, as well as any text or markings around the border of the map.

Figures 6.1 and 6.2 show maps with surrounding background. Figure 6.3 shows a typical map with no surround background.

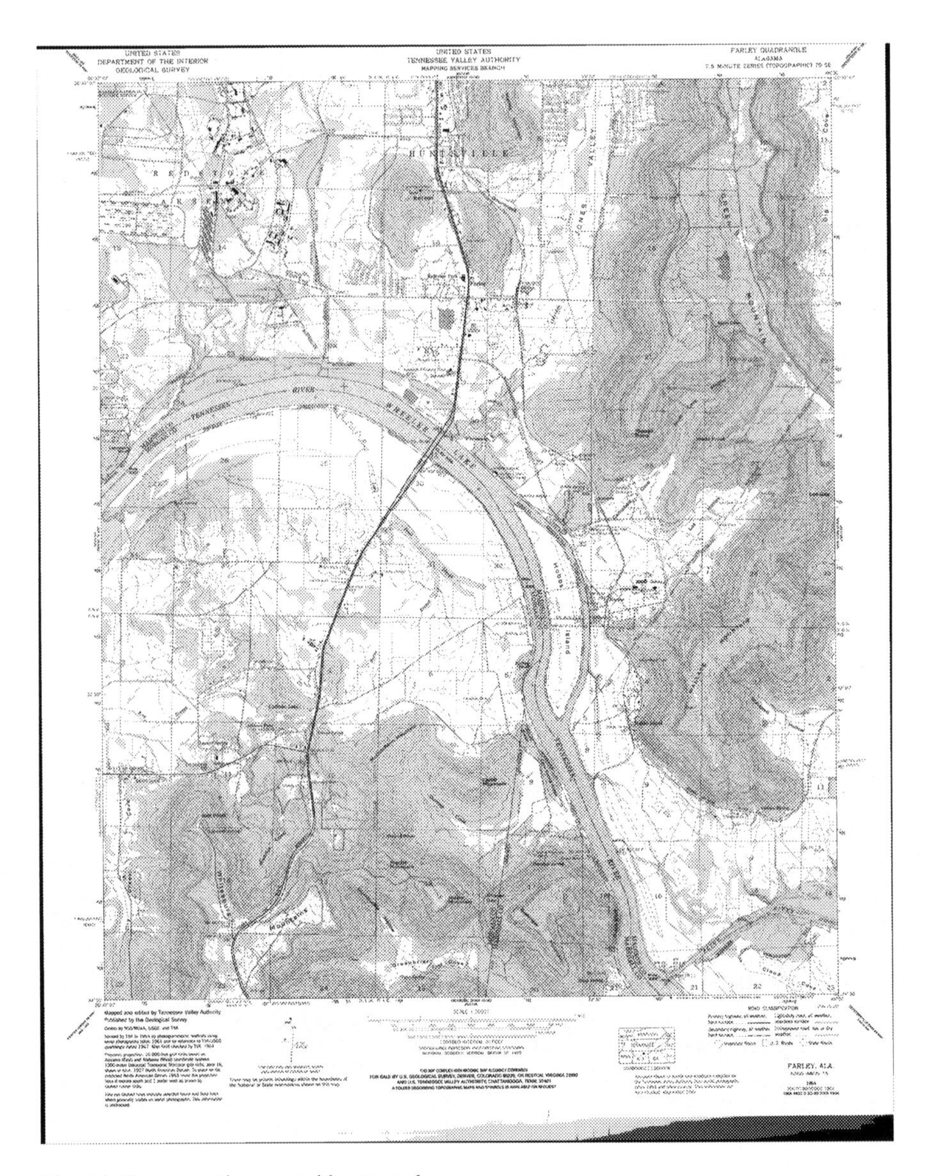

Fig. 6.1 Representative map with surround

6.2.1.1 Map Surround Background Segmentation

Several methods were tested to perform *Map Surround Background* segmentation. One approach is to make use of the connected components of the **WHITE** pixels. This has some strong aspects (e.g., usually produces very accurate results), but has some serious drawbacks: it is very slow and has major problems if the surrounding

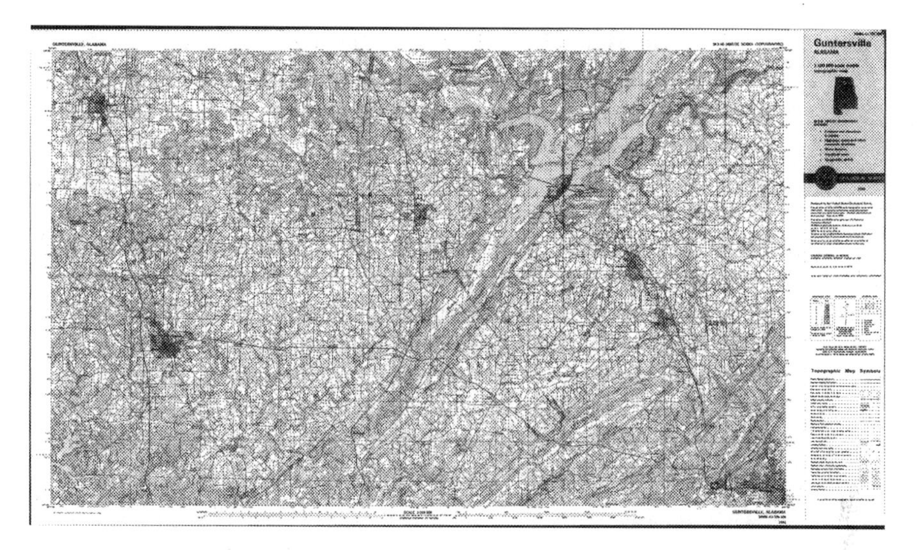

Fig. 6.2 Another representative map with surround

Fig. 6.3 Representative map with no surround

background connects into the map proper. The method used here is to project the sum of the **WHITE** pixels in the horizontal and vertical directions and detect a large drop in the function values going from each end; this is described in more detail below. Other approaches were also considered (e.g., finding the map corners, or finding the black border of the map), but the projection method works very well so these other alternatives were not explored in much detail.

Connected Component Method (CC Method) The CC based map segmentation worked well on the 3 tests, but took hours to run. The basic idea was to calculate the area of the rectangle enclosing each CC, and take the CC with the largest area as the map area.

The more efficient version takes the middle ten columns of the image, and the middle ten rows of the image, and determines the most prevalent CC. Then this CC is extracted from the image. This works well and is very fast. Interior regions that are not part of the map CC are filled in with Matlab *imfill* function.

Tests were run on:

- maps with significant surround: e.g., $O34086E5$.
- maps with no surround: e.g., $q1120\_drg24$.

There are some problems: if the CC is not completely connected around the boundary, then the surround background will extend into the map area. This is overcome by looking for pixels in the map area such that their immediate row+1 neighbor down one pixel is background and there exist non-background pixels somewhere lower down the column. This is filled in. Once this is done, than a final round of *imfill* is used to fill in any interior holes. However, this is very slow computationally.

Projection Method The projection method consists of the following steps:

- Compute the number of **WHITE** pixels in each column; call this f_V.
- Compute the number of **WHITE** pixels in each row; call this f_H.
- Track in f_V from column 1 in increasing column value and determine if there is a large drop in **WHITE** pixel count. If so, mark that as the left map boundary.
- Track in f_V from the last column in decreasing column value, and determine if there is a large drop in **WHITE** pixel count. If so, mark that as the right map boundary. (Note that in both these tracking steps, it is necessary to get past any legend area that may look similar to the map border.)
- Rotate the map image 90° and repeat the above 2 steps to find the top and bottom boundaries.
- If the projection of **WHITE** indicates that there is a surround, then use the projections to find the left, right, top and bottom limits of the surround; form the map area this way and return.
- Else make the map area fill the image area, and surround area empty.

Figure 6.4 shows the extracted map area (in white) and surround (in black) for the map shown in Fig. 6.1 (O34086E5). Tests were conducted on a large number of images, and the results were excellent.

Fig. 6.4 Extracted central map area from map with surround (O34086E5)

6.2.2 Map Embedded Background

For general maps, the background pixels need to be determined from an analysis of the color usage in the image. For USGS maps, the *Map Embedded Background* is generally comprised of the color **WHITE**. In the extraction of many features (other than vegetation), it is also useful to consider the color **GREEN** (index value equals 5) as background. Figure 6.5 shows a subwindow of $F34086A1$ and the Fig. 6.6 shows its foreground pixels.

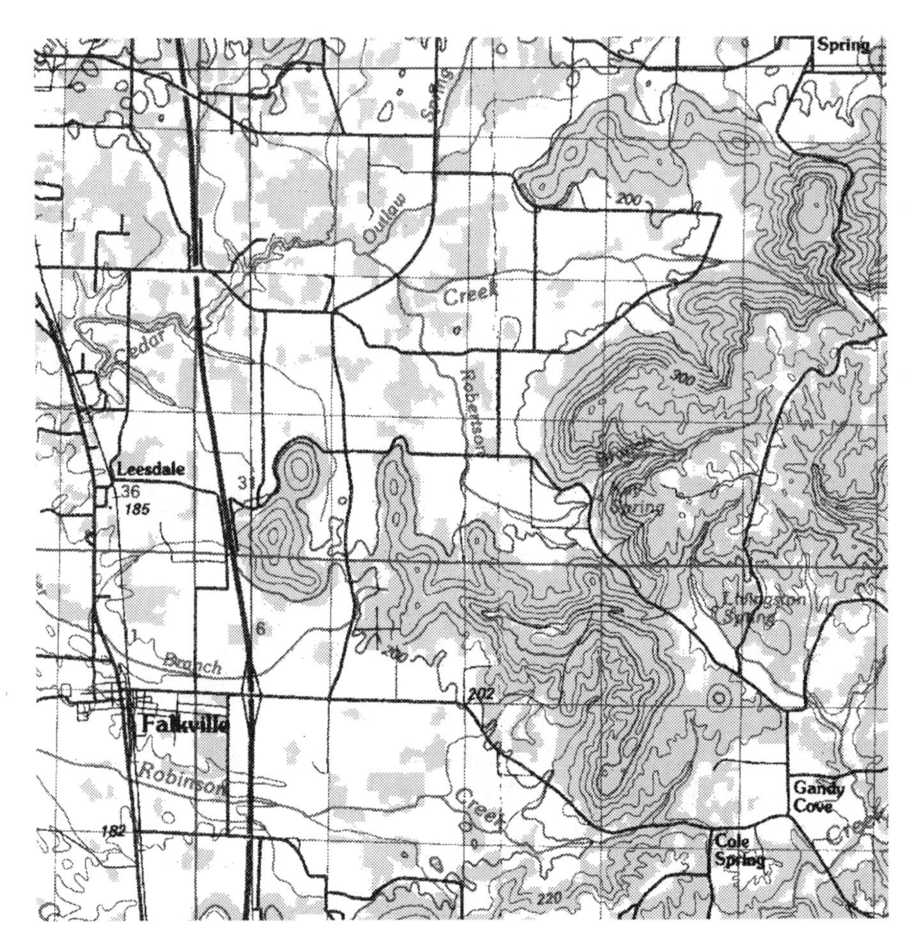

Fig. 6.5 Subimage of F34086A1

Fig. 6.6 Foreground of subimage of F34086A1

Chapter 7
Road and Road Intersection Extraction

7.1 Introduction

Beginning[1] in 1879 the United States Geological Survey (USGS) began surveying land in the United States. Since then they have developed over 55,000 1: 24,000-scale topographic maps covering the 48 coterminous states in a standard, detailed manner. The result is a wealth of data contained in physical documents. Unfortunately many of these documents over the years have begun to deteriorate.

Advances in digital information technology of the twenty-first century have brought about the conversion of these older physical documents, into digitized representations (see Fig. 7.1). The knowledge available in these maps is important and therefore it is critical to be able to recover the semantic content (roads, iso-contours, road intersections) of these into meaningful, accurate representations. New methods are developed here in order to extract higher order semantic content.

Work on extracting these features has already started; however, parsing the document to extract semantic content has proven difficult. The documents themselves have been scanned and color corrected, with the result that the digitization of older maps has introduced noise and error in the digital versions. Overlapping features and gaps in the data cause existing GIS extraction tools to fail. Along with these issues, the amount of descriptive information and variety of symbols have compounded the problem.

To overcome this a framework is developed which implements a robust set of organizing rules used in graphic design and psychology derived from fundamental principles of Gestalt perception. Gestalt principles are derived from the Law of Prägnanz which defines a philosophical method for segmenting objects and an explanation of human perception. The Gestalt principles include the law of closure, similarity, proximity, symmetry, continuity and common fate. Of these laws we

[1]This chapter is contributed by Trevor Linton based on his MS thesis.

T.C. Henderson, *Analysis of Engineering Drawings and Raster Map Images*,

DOI 10.1007/978-1-4419-8167-7_7, © Springer Science+Business Media New York 2014

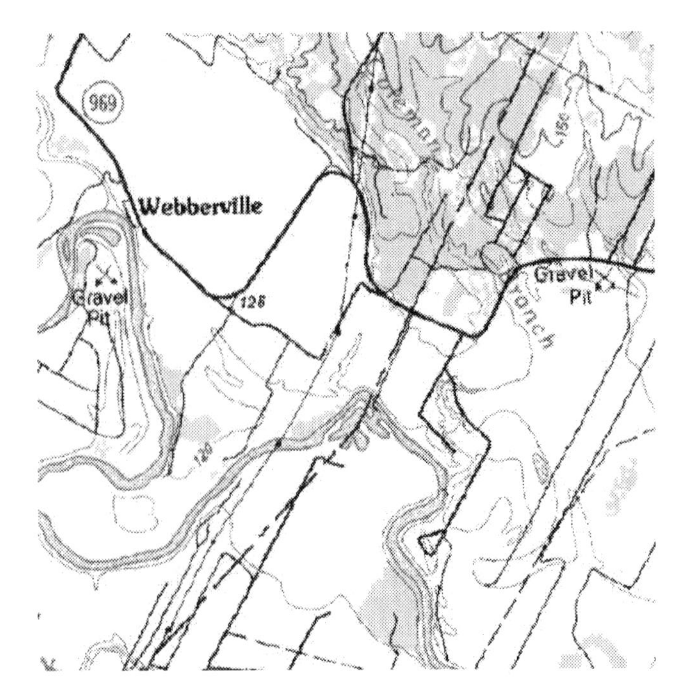

Fig. 7.1 USGS sub-image sample map

attempt to implement similarity, proximity and continuity using various image processing techniques.

The principles of similarity, proximity and continuity are important as they allow the consideration of different perspectives in reconstructing features from noisy data. For a curve segment with gaps, an algorithm implementing the principle of continuity helps correct these errors. In addition, an algorithm which adheres to the principle of similarity such as a histogram model analysis can be very effective at creating rough estimates of features.

Such a method requires the extraction of lower level features such as curves, lines, regions, boundaries, outliers, symbols and junctions as the basis for analysis. The extraction of complete and accurate representations of lower order features helps to derive higher order semantic features such as interstate and state highways, rivers, etc. We describe methods for extracting low level features from raster images containing geographical map data, and for obtaining robust segmentations of semantic features in the map.

The road framework is broken up into a pre-processing stage, tensor voting and a post-processing stage shown in Fig. 7.2. Using methods such as histogram model analysis, dilation, erosion and thinning has proved adequate at creating an initial estimate of lower level features in the maps. Using tensor voting afterwards we can then fill in and clean the image by reinforcing Gestalt principles.

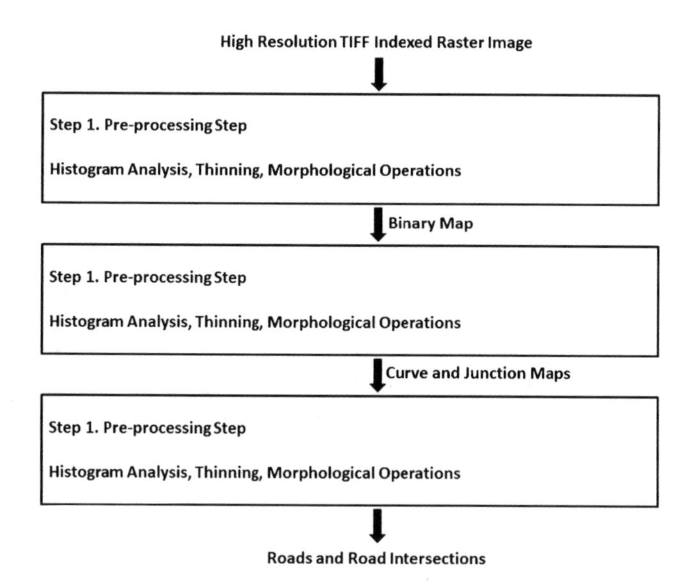

Fig. 7.2 Overview of the road framework

The resulting output of the tensor voting system is further processed in order to create objects that can be classified and used in order to extract higher order features. Connected components, thresholding, thinning, and local maxima are used in this post-processing phase and are compared and contrasted for their effectiveness. The combined efforts of this process allow us to achieve noise reduction, fitting, junction analysis, gap filling and region boundary extraction.

Tools and models developed in the framework are also computationally effective with respect to their memory and CPU time footprint and performance. This allows useful exploitation in actual data analysis settings.

7.1.1 Goals

The goals of this work arise from the requirements of real life use of extracted content. The primary use of the results from this research is to register the semantic content with other GIS (Geographic Information Systems) software. In order to achieve this, we must determine the primary needs of GIS software in order to exploit the results. The primary features considered are roads and road intersections; however, this method can be expanded to other linear features with different histogram models.

The detection and localization of intersections provide a clean quantitative framework for the analysis of performance and is highly useful; e.g., to register maps with aerial images. Using more complicated objects such as curve segments requires more extensive analysis with ground truth data.

$$Recall = \frac{|\,relevant \cap retrieved\,|}{|\,relevant\,|}$$

$$Precision = \frac{|\,relevant \cap retrieved\,|}{|\,retrieved\,|} \tag{7.1}$$

Goals are set and performance evaluation for road intersection detection is performed quantitatively through the recall and precision measures defined in Eq. (7.1) (*relevant* is the set of relevant pixels in the image, while *retrieved* is the set of pixels returned by the algorithm). Any two corresponding intersection points are matched if they are within a specified distance (in pixels) of each other in the raster map. Quantitative analysis of linear curve segments is done by comparing the linear curve segments to ground truth images. In situations of gap filling or fitting an optimal curve, manually generated ground truth is used as a comparison for the missing data.

Production use of the framework must also consider runtime performance and memory or CPU-time footprints and constrain solutions to algorithms that can be run in a realistic amount of time. The tensor voting method specifically can run in a completely parallel fashion and thus constant time (if enough resources are available).

The major goals here are:

- To develop road segmentation algorithms for raster maps that perform at the level of 95% recall and 95% precision for roads in a quantitative analysis.
- To develop road intersection detection algorithms for raster maps that perform at the level of 99% recall and 90% precision for intersections in a quantitative analysis.

7.2 Related Work

Work on extracting semantic content from raster maps has faced many challenges. Maps are scanned under an extremely high resolution TIFF format. A lot of work has been done on color correction and accurate conversion of the scanned data into an indexed color palette; removing color distortions due to scanning was once a topic of research but now has been adequately addressed [63, 94].

With high resolution images and reasonably accurate color in map images, the subject of research is now the extraction of semantic data. Recent algorithms in the subject of topographical feature extraction typically use a system of thinning to produce vector representations of linear features, followed by the application of the A* algorithm to connect components along with some sort of preference and simple pattern matching, geometric analysis or histogram analysis to determine features in the raster map image.

The framework described utilizes a method called tensor voting proposed by Medioni [84, 85]. The method is useful for de-noising and reconstructing linear features from a sparse data set. The implementation of tensor voting is based on the description in [84].

The method used for implementing tensor voting in the USGS maps is similar to that described by Shao et al. [105]. The method proposed for implementing tensor voting and systems for pre-processing and post-processing involves thresholding, thinning and de-noising the output of the curve map from the tensor voting system.

Chiang et al. [135] describe a method for extracting roads, filling gaps in roads and identifying intersections. The method extracts pixels thought to be roads using parallel line tracing techniques as well as simply extracting layers of color which correspond to road. The road data is then dilated, eroded and thinned to fill in gaps. The result of this is a 1 pixel wide (unit width) road that was examined for intersections by looking for pixels with three or more adjacent connected pixels. This method for filling gaps may not work for larger gaps as well as situations where gaps curve. The accuracy due to thinning may not keep the original road boundaries intact. Their results from USGS topographic maps for intersections had a precision of 84% with a recall of 75% and a positional accuracy of 80%. A similar method of thinning is used in the pre-processing steps but further processing is needed in order to fill gaps and obtain better performance.

Khotanzad and Zink [64] propose a method for contour extraction which first removes any colors not part of the feature set to be examined. Linear features are extracted using valley seeking algorithms, then the A* algorithm is used to connect valleys together and form linear features or close gaps. While this method has been studied in depth, A* is not as effective at following a curve due to its tendency to prefer proximity over continuity which leads to connections that are not correct. Another concern is the time and space complexities of A*, although these might be overcome with optimization, thinning techniques and the right heuristic function built into the A* algorithm. The results of this method on the sample images in the paper were qualitatively excellent, but quantitatively had incorrectly classified 1.5% of the non-contour lines as contour lines, and 2.4% of contour lines were misclassified as other features. While this method is effective, its results on roads rather than contours is still unknown and it is not used to perform intersection detection.

Ahn et al. [8] give a method of color separation, noise elimination using erosion and dilation, and thinning and vectorization. This method was used on Korean topographical maps and seemed effective; however no quantitative results were given. The method also seems to be unable to overcome severe noise and gaps in linear features.

The pre-processing techniques described here utilizing histogram models to find feature estimates are based on work published by Henderson et al. [46]. The models formed are capable of finding good estimates of features that can be later used for tensor voting and post-processing methods.

Pouderoux and Spinello [96] propose a method to reinforce and fill gaps in contour lines using the Gestalt principle of good continuation. The method is based on a gradient orientation field generated by the contour lines and uses the tangent at the ends of the gap to approximate the best curve between the two points without crossing another linear feature. The framework described uses similar principles to our method although we use tensor voting to enforce good continuation rather than a gradient orientation field. An example was given in the paper; however, no quantitative results were published.

Pezeshk and Tutwiler [94] give a system of histogram equalization (HE, AHE and CLAHE) to color correct USGS images and enhance features in the image. They did not publish quantitative results for this method; however, enhancing a feature through blurring, dilation and subsequently a form of histogram equalization may prove useful for enhancing a feature prior to extraction.

Miyoshi et al. [86] provide a review of methods for extracting buildings utilizing geometric features. Extraction was achieved by thinning line segments and using a vector-chaining procedure to produce various types of connected, branching and non-connected vectors. Features of the vectors were used to identify buildings and other features in the image. For example, a connected loop of a certain size would be classified as a building while any non-connecting loop or loops which were too large or did not contain sides with a straight line were not considered buildings. The results of this method correctly identified 87.3% of village buildings, 83.3% of urban buildings and 89.4% of residential buildings. False positive identifications were respectively 5.2%, 3.3% and 9.7%.

Using methods of this type may be effective; however, the data input used in these examples were well formed scanned drawings. The USGS data set used in this thesis contains various amounts of noise and gaps in the data; making preprocessing steps useful to extract lower order features.

The analysis of technical and engineering drawings using distributed agent based systems has been shown beneficial. Swaminathan [113] and Henderson [48–50] examined this technique using various image processing techniques combined with an agent based system and a hierarchy of grammars to examine if communication of networks of agents could better correct for errors and identify both syntactic and semantic symbols in drawings (see Chap. 5). The use of agents showed a significant ability to account for unknown semantics in engineering drawings.

A similar method of thinning and vectoring were described by San, Yatim et al. [104]. Connecting components of contour lines was done using radial symmetry in the contour lines and closest extremity. The initial problem with this approach is the lack of smooth curves between disconnected components.

Zheng et al. describe a method [140] for extracting roads out of satellite images. The method uses colors to extract specific road types and make an initial estimate of the feature. The tensor voting framework was then used on the binary image to produce a curve map. The tensor voting framework curve map was thresholded and then thinned in order to make a unit width vector of the road. The results were qualitatively good however no quantitative results were published. The method used here is similar; however, the pre-processing and post-processing techniques are different. The method described by Zheng et al. was applied to satellite images while the method described here is applied to USGS maps.

Poullis et al. [97] used a similar method of tensor voting on satellite image data, but combined it with other images such as aerial photographs to more accurately segment the image. No quantitative results were reported.

Hinz and Baumgartner [54] describe a method of identifying roads in aerial images by first segmenting the image into urban and rural areas, removing shadows, occlusion and building outlines. The result is then segmented using histogram model analysis, thresholding and then modeling techniques. The result of this method is 75% recall and 95% precision for roads.

Li et al. [69] published a method for extracting labels, roads and graphics from USGS maps. The method traces roads until text is found and wraps it in a bound box extracting the text then filling in the gap for the next road. This method was effective at finding well defined roads and text, but failed on curved roads and noisy data.

Podlasovet al. [95] give a method of restoration of binary semantic layers of map images. In the paper they proposed a method for restoring raster image maps from artifacts caused by color separation. However, their method was typically used on maps which are in a RGB or CMYK color space, whereas we work on images with color indexed palettes that are well defined.

Ageenko and Podlasov [6] propose a novel method for removing noise and reducing error in features. The method to extract a feature begins by taking a union of overlapping features and then dilating the feature of interest and using the unioned features as a mask to reduce the affects of over-dilation. The results of the method were taken over NLS Map Series for the water and field features. The results reduced the error rate in the features by 12.72% to 14.14% for water and fields, respectively.

Chaing et al. [21] give methods for extracting roads and intersections using a wide variety of image processing techniques from thinning, erosion, dilation and histogram analysis. The use of double-line format detection or parallel-pattern tracing was used to identify roads via a geometric analysis of the structure in the map. A knowledge based logical process was followed to extract the roads and intersections once the pre-processing steps were finished. The method described has a precision of 82% and a recall of 60% for USGS Topographic maps.

Previous work with tensor voting and post-processing methods has been described by Henderson and Linton [45]. The methods introduced there form the basis for the work described here. The preliminary results published in that paper are 93% recall and 66% precision for road intersections.

7.3 Approach

7.3.1 Overview

To extract features from USGS maps, a system is built utilizing Gestalt principles of similarity, continuity and proximity. The system is separated into three parts: a pre-processing phase, tensor voting and a post-processing phase. The pre-processing

Table 7.1 USGS index to RGB values

Indexed pixel value	Color	R value	G value	B value
0	Black	0	0	0
1	White	255	255	255
2	Blue	0	151	164
3	Red	203	0	23
4	Brown	131	66	37
5	Green	201	234	157
6	Purple	137	51	128
7	Yellow	255	234	0
8	Light Blue	167	226	226
9	Light Red	255	184	184
10	Light Purple	218	179	214
11	Light Gray	209	209	209
12	Light Brown	207	164	142

phase cleans up initial noise, finds rough estimates of features utilizing histogram models and reduces features to smaller unit width through thinning. The tensor voting phase reduces noise, bridges gaps and reinforces curves in a manner which preserves continuation over proximity. The post-processing phase takes the resulting output from the tensor voting system, thresholds the data and then uses a variety of techniques discussed here to accurately extract unit width vector features.

The result of this process is a binary unit width segmentation of the specific linear feature of interest. Processing on these segments then allows us to extract further details such as intersections (junctions) within the linear features.

7.3.2 USGS Maps

United States Geographical Survey maps are large (average of 9495 by 5552 pixels and 15 megabytes) TIFF 8-bit color indexed raster images. Care is taken to make sure the maps are read in and not converted to any other color-space or format. Casting the raster map into a different color-space could result in inaccurate colors due to interpolation. Converting the map into a lossy image format may result in compression artifacts and inaccurate colors and geometrical features.

The maps contain a range of features including (but not limited to) contour lines, geopolitical boundaries, symbols, roads, train tracks, labels, lakes, rivers, highways and freeways. While most maps only contain a small number of colors, features are identified through mixed color combinations giving a distinct texture to a feature. Human defined symbols like text and road markers can be found by exploiting their distinct color and shape (Table 7.1).

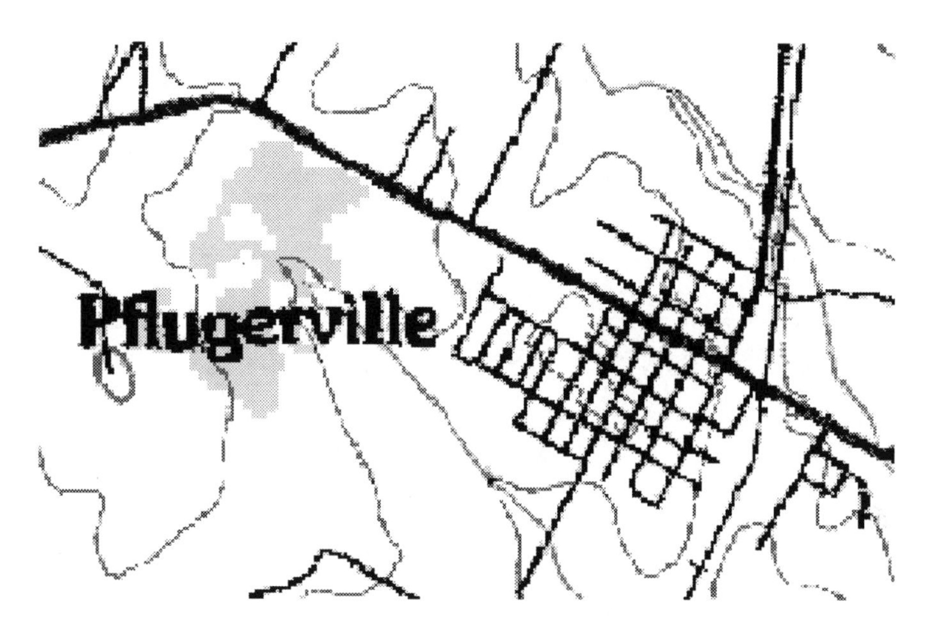

Fig. 7.3 USGS Sub-image of a city

Cities are generally identified with smaller clusters of roads and a text label identifying the name of the region such as in Fig. 7.3. Contour lines are generally identified with a brown color and are condensed curves (Fig. 7.4) describing topographical information and height of the area. Symbols in maps such as in Fig. 7.5 identify contextual information for roads, cities and other geopolitical information. Large areas with a dotted blue and white pattern identify water such as in Fig. 7.6. We limit ourselves to the extraction of roads, usually red and black depending on the type of roads, which are generally straight intersecting lines are shown in Figs. 7.3 and 7.7.

Roads can be defined by a variety of textured colors. Some maps include a pink or black double lined roads with specific markers for direction. Others include thicker black and red lines to describe freeways, thinner to describe highways and solid thin black lines to describe regular roads, urban or rural. Train tracks and other railways may be described by a black line with stylized cross ties.

Features may overlap in the map making gap filling a necessity in order to accurately extract and represent roads, contour lines, etc. Gaps may be exceedingly large which represents a challenge in accurately estimating a curve or line correctly. A larger problem arises when two distinct features having similar textures overlap. This can make correctly segmenting features increasingly difficult.

Text is largely ignored in this analysis as document analysis has defined methods for extracting text within maps. The process of extracting text is made simpler as the need to accurately represent the original shape of the text (as is necessary with other features such as roads) is not required.

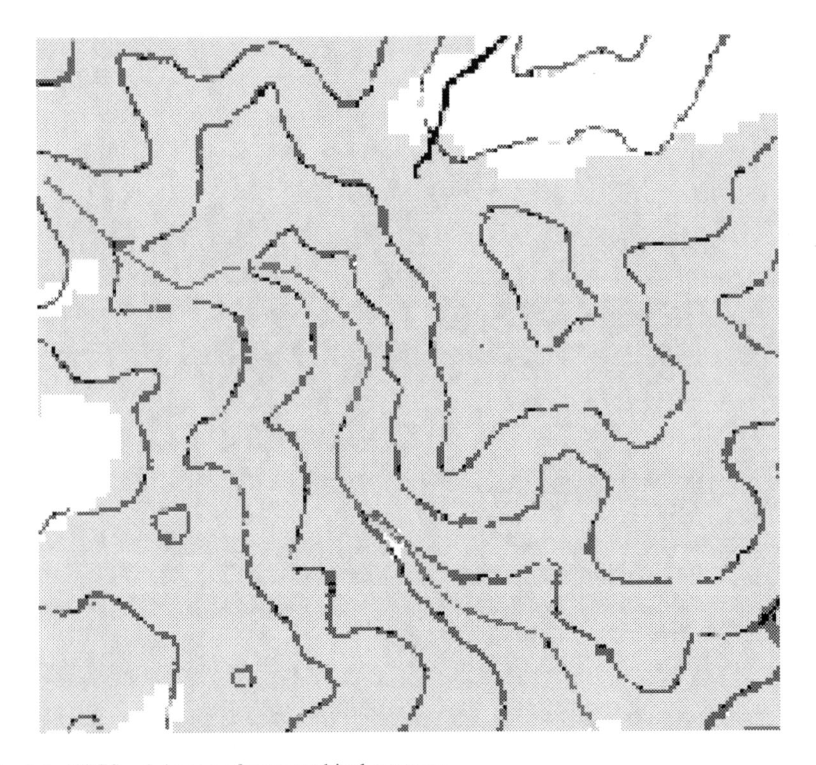

Fig. 7.4 USGS sub-image of topographical contours

Fig. 7.5 USGS example of a symbol in a map

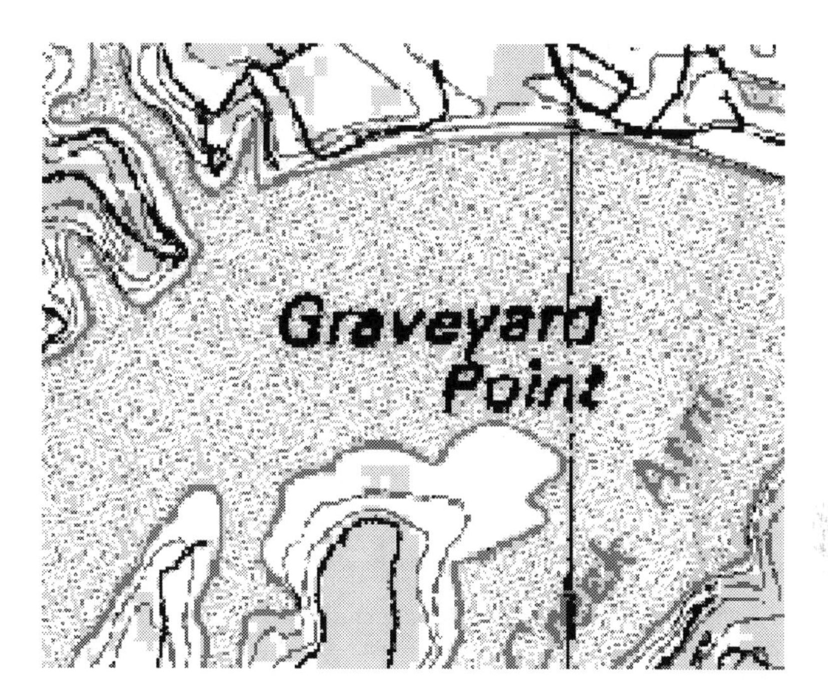

Fig. 7.6 USGS sub-image of rivers and lakes

Fig. 7.7 USGS sub-image of roads and freeways

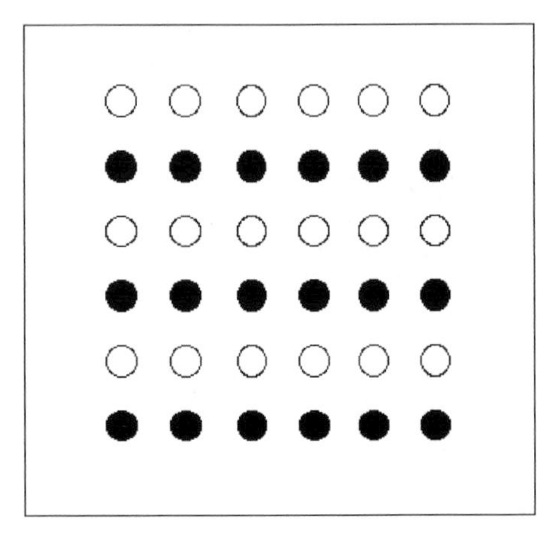

Fig. 7.8 Demonstration of the law of similarity

7.3.3 *Gestalt Principles*

Gestalt psychology or gestaltism is a theory of the brain's self-organizing capability with respect to visual recognition of figures and shapes. Within Gestalt psychology the Law of Prägnanz states that we experience visual phenomena as regular, orderly, symmetric, and simple. In an attempt to refine the Law of Prägnanz, Gestalt psychologists have created the Gestalt principles which help organize visual information. These principles include the law of closure, similarity, proximity, symmetry, continuity and common fate. In an attempt to achieve image segmentation, some of these laws or principles are used in our framework in order to draw out features.

- *Similarity*: The grouping of similar elements into a whole element or grouping based on shape and texture to complete a regular figure or region. As can be seen in Fig. 7.8 the dark circles and outlined circles produce separate segmentations in the image.
- *Proximity*: The amount of space between objects and pixels is a tool for grouping items together and determining boundaries, regions and features. As can be seen in Fig. 7.9 the spacing between the lines of circles on the left cause a segmentation.
- *Continuity*: Patterns which repeat are followed by the mind, even in the event the pattern is missing in places, the mind will still fill in gaps and conjure up the complete pattern in the mind. The principle of continuity is demonstrated in Fig. 7.10.
- *Closure*: The mind is drawn towards regularity removing outliers and perceives items with a complete and regular figure. The principle of closure is illustrated in Fig. 7.10.

These principles are represented in the methods used here. The law of similarity is exploited through the use of histogram models. The histogram models match various textures and group them together to get rough estimates of features. The laws

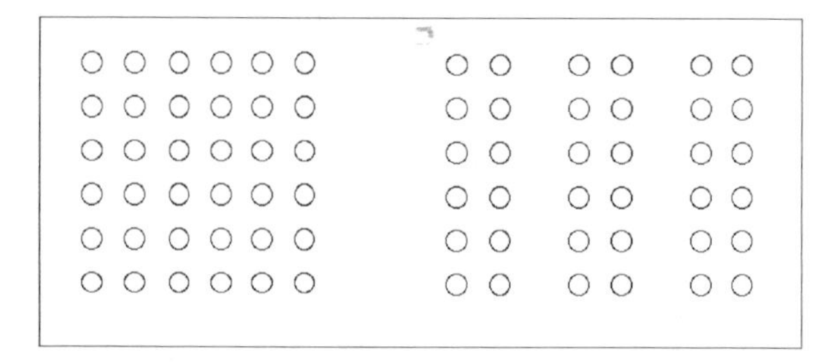

Fig. 7.9 Demonstration of the law of proximity

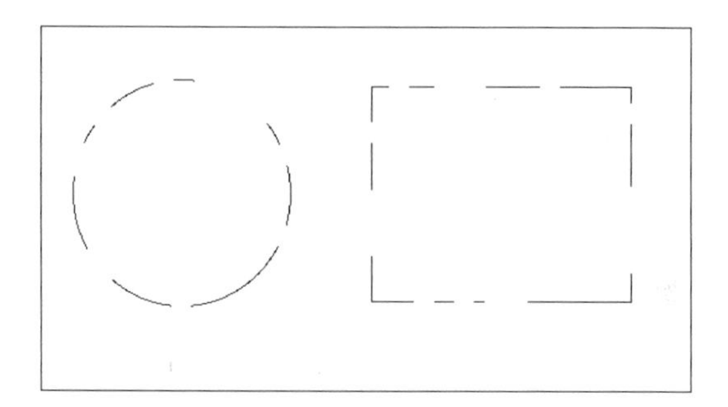

Fig. 7.10 Demonstration of the laws of closure and continuity

of proximity and continuity are represented through the tensor voting system. The tensor voting system enforces good continuity and proximity when filling gaps and optimally finding curves. The law of closure is expressed through the tensor voting system and the post processing methods.

7.3.4 Low Level Features

In order to extract semantic features low level features must first be analyzed. Low level features include lines, curves, color histograms, junctions, regions and end points. The four main low level features to be exploited are:

- Linear nature of the pixel (lines, curves).
- Color histogram of local window at each point.
- Pixel regions.
- End points.

There are a few constraints on the output of lower order features:

1. Each feature must be independent from others. For example two crossing roads should be separated into the two road segments and an intersection as low level features.
2. Segments or objects must be unit width and connected.

Fig. 7.11 Initial image (*left*), initial estimate of features (*middle*), cleaned up initial estimate (*right*)

3. Gaps must be filled with the closest approximated curve and in a manner which preserves continuity, not proximity.
4. Features must be semantically smooth and preserve the original feature in the raster image.

7.4 Pre-processing Techniques

To properly prepare the raster maps for analysis by the tensor voting framework, a series of pre-processing steps is necessary to remove noise and approximate feature classes. The process begins by assigning initial feature labels using a variety of techniques. This segmentation produces a binary image. The results are then cleaned by removing outliers and isolated elements. The data is then run through a skeletonization process as seen in Fig. 7.11. Once properly cleaned the results are then used in the tensor voting framework.

7.4.1 Histogram Models

USGS maps have a well-defined structure which we can exploit. Using the Gestalt principles of similarity, initial assignments can be made on the various types of roads in the map by examining histogram models representing a class. In order to make initial assignments of pixels, the specified features' histogram model is created as a set of sample histograms representative of the class. The samples are taken from areas in the maps where each feature is shown in Fig. 7.12. Each class has the same number of histogram samples in order to prevent bias to any one class. Tables 7.2–7.5 are common road types and their corresponding histogram models.

Most features in the system are linear and can be analyzed with a histogram model as their geometric structure is less important than color texture; this makes histogram models effective. However some features such as route markers have a geometric shape which makes it necessary to employ some other detection system for them.

Fig. 7.12 Example of USGS symbols legend

Topographic Map Symbols

Primary highway, hard surface
Secondary highway, hard surface
Light duty road, principal street, hard or improved surface
Other road or street; trail
Route marker: Interstate; U. S.; State
Railroad: standard gage; narrow gage
Bridge; overpass; underpass
Tunnel: road; railroad
Built up area; locality; elevation
Airport; landing field; landing strip
National boundary
State boundary
County boundary
National or State reservation boundary
Land grant boundary
U. S. public lands survey: range, township; section
Range, township; section line: protracted
Power transmission line; pipeline
Dam; dam with lock
Cemetery; building
Windmill; water well; spring
Mine shaft; adit or cave; mine, quarry; gravel pit
Campground; picnic area; U. S. location monument
Ruins; cliff dwelling
Distorted surface: strip mine, lava; sand
Contours: index; intermediate; supplementary
Bathymetric contours: index; intermediate
Stream, lake: perennial; intermittent
Rapids, large and small; falls, large and small
Area to be submerged; marsh, swamp
Land subject to controlled inundation; woodland
Scrub; mangrove
Orchard; vineyard

Table 7.2 Primary highways

Indexed pixel value	Color	Number in sub-image
0	Black	1030
1	White	2,925
2	Blue	7
3	Red	456
4	Brown	352
5–12	Green	0

Several different methods are used to create initial assignments of features. A process of using mean, standard deviation and other statistical methods called the class conditional density with Mahalanobis distance is shown using the histogram models described earlier. A knowledge based classifier uses information known about the various features to identify and classify them. This method seems to work well at producing sharp and clean assignments with minimal dilation or

Table 7.3 Secondary highways

Indexed pixel value	Color	Number in sub-image
0	Black	533
1	White	3,300
2	Blue	1
3	Red	366
4	Brown	305
5–12	Green	0

Table 7.4 Light duty road

Indexed pixel value	Color	Number in sub-image
0	Black	807
1	White	2,404
2	Blue	111
3	Red	0
4	Brown	12
5–12	Green	20

Table 7.5 Other street

Indexed pixel value	Color	Number in sub-image
0	Black	163
1	White	1,060
2	Blue	45
3	Red	0
4	Brown	8
5–12	Green	0

bleeding. The last process considered here is the k-nearest neighbors method using the histogram example as the model described earlier. The methods described here are compared qualitatively with respect to the map shown in Fig. 7.13.

7.4.2 Class Conditional Density with Mahalanobis Distance

The class conditional density (CCD)with Mahalanobis distance classifier is a statistical method to classify each pixel in the image. The class model is the mean of the histogram examples defined by Eq. (7.2) and the variance of each class of histograms defined by Eq. (7.3). The Mahalanobis distance is defined by Eq. (7.4). Eq. (7.4) is calculated for each k and the k with the minimal value is the assigned class. The result of this method on a 500 by 500 pixel region is shown in Fig. 7.14.

$$\overline{\mu}_k = \frac{1}{N_k}\sum_{i=1}^{N_k}\overline{x}_{k,i} \tag{7.2}$$

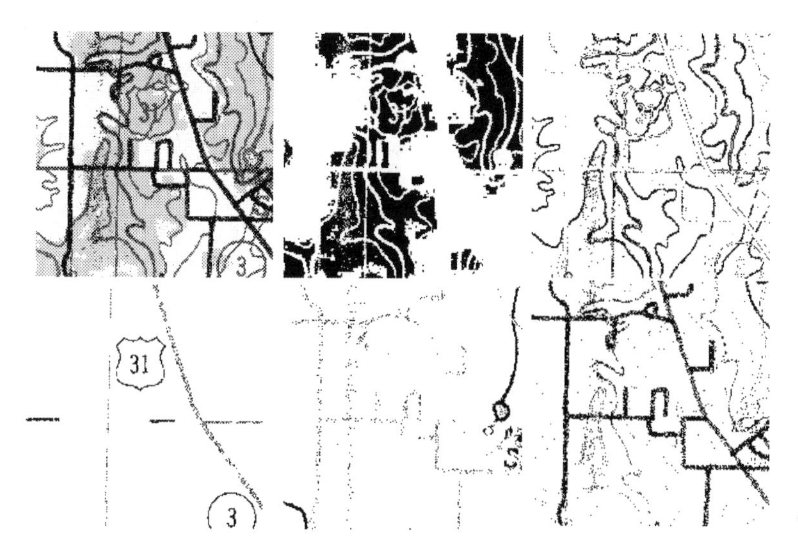

Fig. 7.13 Original image used in comparing the various pre-processing methods below. Original (*top left*), forest (*top middle*), contours (*top right*), geopolitical symbols (*bottom left*), water (*bottom middle*), roads (*bottom right*)

Fig. 7.14 Roads classified by CCD with Mahalanobis distance

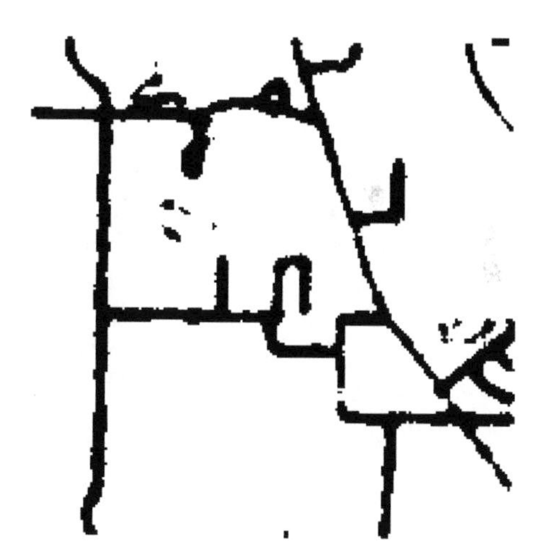

$$\Sigma_k = \frac{1}{N_k - 1} \sum_{i=1}^{N_k} (\overline{x}_{k,i} - \overline{\mu}_k)(\overline{x}_{k,i} - \overline{\mu}_k)^T \tag{7.3}$$

$$\delta(\overline{x}; \overline{\mu}_k, \Sigma_k) = \frac{1}{2} \sqrt{(\overline{x} - \overline{\mu}_k)^T \Sigma_k^{-1} (\overline{x} - \overline{\mu}_k)} \tag{7.4}$$

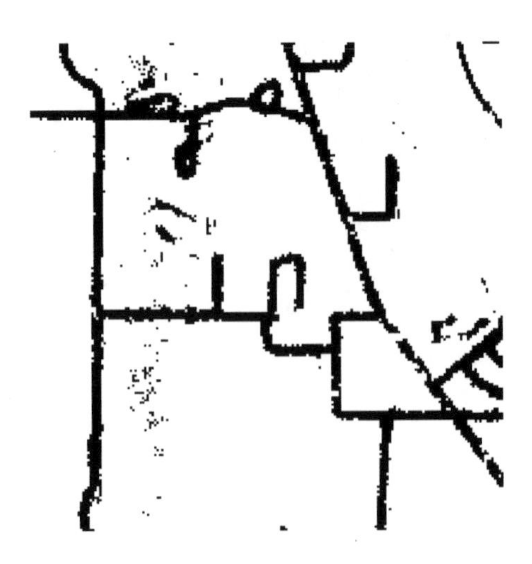

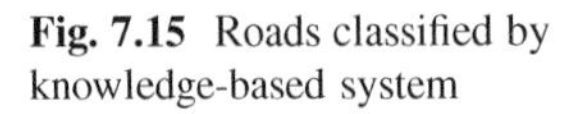

Fig. 7.15 Roads classified by knowledge-based system

7.4.3 Knowledge-Based Classifier

To classify features in the image, a knowledge based approach can be used in which the user specifies a predicate which characterizes class membership. The approach is based on a normalized histogram of an area around the pixel that will be classified. The process uses the normalized histogram bins to determine the class. The results of a road classification are shown in Fig. 7.15.

To classify roads the bins of the histogram are analyzed and roads are further sub-classified into light duty and primary roads. The light duty roads are identified if the color black is the largest color in the histogram, there is blue in the histogram, there is less or equal amounts of red than blue, and there is less brown than blue. Primary roads are classified if black is the most numerous color in the histogram, the color blue is less than or equal to red, blue is less than or equal to brown, there is some red in the histogram, and there is some brown.

7.4.4 k-Nearest Neighbors

The k-Nearest Neighbors process begins by taking a training set of examples (histograms at selected pixels and their respective classes) and treating these as vectors, producing an n-dimension vector where n is the number of bins in the histogram. The histogram from the window surrounding the point of interest is then represented in the vector space and the k closest neighbors to it are found. The distance is calculated by the Euclidean distance defined by Eq. (7.5) where p is the point of interest and q is the point of the histogram model and the index i

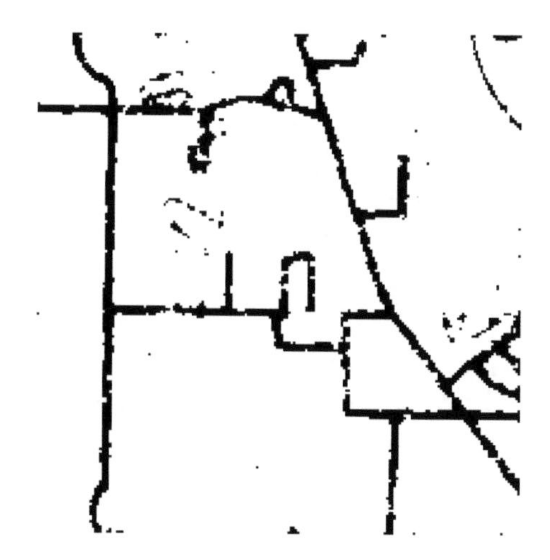

Fig. 7.16 Roads classified by k-nearest neighbors

represents the bin. The assignment is then determined by finding the class out of the k neighbors that is most frequent. An example of roads classified by k-nearest neighbors is shown in Fig. 7.16.

$$D = \sqrt{\sum_{i=1}^{n} (p_i - q_i)^2} \tag{7.5}$$

7.5 Tensor Voting

Each pixel classified as road during the pre-processing step gives rise to a curve estimate through that pixel; this curve is then used to cast votes at other pixels where the curve should pass. A non-road pixel will thus accumulate votes for the various curves that could pass through it, with straight continuations favored over those sharply changing direction. Thus, a closer curve may have less influence than a more distant one for which the pixel provides straight continuation. Linear features that are in close proximity but have little continuation may still be connected if no other ideal curve can be found.

Tensor voting is a powerful image processing tool that helps clean images, fill gaps and reinforce curves based on Gestalt principles of closure and continuity. The tensor voting system takes in a binary input image and outputs a junction and curve map (see example shown in Fig. 7.17). The curve map gives the likelihood of the presence of a curve passing through each pixel. Likewise the junction map expresses the likelihood of a junction at a pixel. The curve and junction maps are valued from 0 to 1. The tensor voting system takes two parameters, σ and c. While

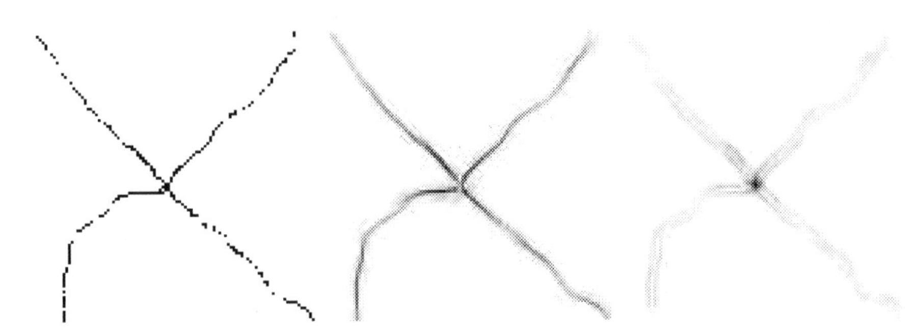

Fig. 7.17 Synthetic image (*left*), calculated curve map (*middle*), and junction map (*right*). *Darker* means higher likelihood of feature

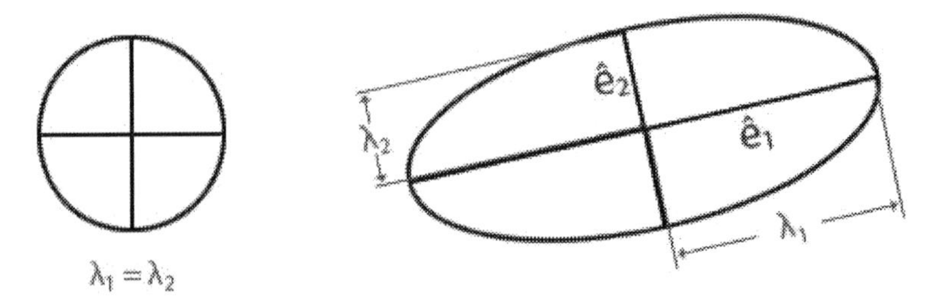

Fig. 7.18 Visualization of a tensor in two dimensions. The *left* is a representation of a junction (*ball*) tensor and the *right* a representation of a curve (*stick*) tensor

c can be derived from σ for optimal balance, it can also be varied to give preference to either continuation or proximity as desired.

Equation (7.6) defines a tensor and is in the form of a second order, symmetric, non-negative definite tensor. The tensor here is a 2x2 matrix and a mathematical representation of the structure type, direction and saliency. The structure type is either a junction or curve and is measured by the difference between λ_1 and λ_2 as shown in Fig. 7.18. If λ_1 is significantly larger than λ_2 the point where the tensor is located is thought to be a part of a curve. If the differences between the two λ's is small, the tensor is thought of as a junction. (Note Medioni referred to the structure types as stick tensors and ball tensors, respectively.) The direction is the alignment of the tensor in the direction of the preferred tangent (or in other words the preferred direction of the curve at that point, where $\hat{e}_1$ is the preferred tangent and $\hat{e}_2$ is the normal.) The saliency is the confidence that this feature exists and is measured by the magnitude of λ_1.

$$T = \lambda_1 \hat{e}_1 \hat{e}_1^T + \lambda_2 \hat{e}_2 \hat{e}_2^T \tag{7.6}$$

The tensor voting process begins by building a sparse tensor field. The sparse tensor field is a PxN array (the same size as the input) with an initial estimate of each tensor at each point defined in the input. The estimates are built using principal component analysis using the direction or estimated tangent as $\hat{e}_1$ and its normal as $\hat{e}_2$. The values of λ_1 and λ_2 are set to one for the initial estimate. Estimates can also be built using a variety of other techniques such as ball voting defined by Medioni or a Canny edge detection algorithm.

Once a sparse tensor field is generated, a voting field is created for each point defined in the sparse tensor field. The voting field is an MxM field where the size (M) is defined by solving for W_{size} in Eq. (7.7). The voting field contains a tensor at each point in the field. The direction of the tensors in the voting field are defined by Eq. (7.8) where θ is the angle from the location of the tensor in the field to the origin (center) of the field defined along the x-axis and l is the distance from the origin to the tensor. The attenuation of the tensors in the voting field is defined by Eq. (7.9) where θ is as defined before, s is the arc length, and k is the curvature. The voting field is aligned with the tangent of the tensor in the sparse tensor field and positioned to be centered above it. Once the voting field is generated and aligned with the tensor in the sparse field, the tensors in the voting field and sparse field are added together to produce a dense tensor field.

$$W_{size} = \sqrt{-\sigma^2 * ln(0.01)} \tag{7.7}$$

$$s = \frac{\theta l}{sin(\theta)}$$

$$k = \frac{2sin(\theta)}{l}$$

$$S_{SO}(l,\theta) = DF(s,k,\theta)\begin{bmatrix} -sin(2\theta) \\ cos(2\theta) \end{bmatrix}\begin{bmatrix} -sin(2\theta) & cos(2\theta) \end{bmatrix} \tag{7.8}$$

$$DF(s,k,\theta) = \left(e^{-\frac{s^2+ck^2}{\sigma^2}}\right) \tag{7.9}$$

The dense tensor field can then be decomposed in order to determine where curves and junctions exist. The field also describes the direction of any curves through the $\hat{e}_1$ value. For each tensor in the dense tensor field the curve map and junction map can be found by considering:

- If $\lambda_1 - \lambda_2 > 0$, then this demonstrates the likelihood of the point being on a curve. The saliency of a stick component being larger than that of the ball component indicates a certainty of the orientation, therefore the tensor most likely belongs to a curve.
- If $\lambda_1 \approx \lambda_2 > 0$, then this demonstrates the likelihood of the point being a junction or irrelevant. If the λ_1 is approximately the same size as λ_2 no orientation is likely and therefore the point can either be considered irrelevant or a junction.

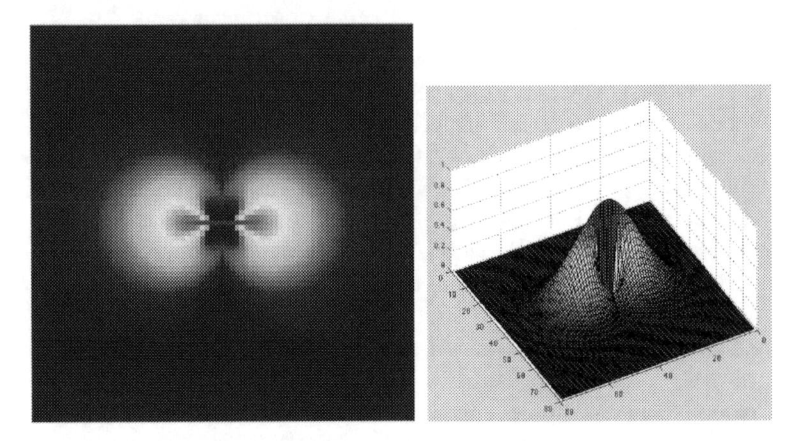

Fig. 7.19 2D and 3D view of a voting field's ideal attenuation as defined by Eq. (7.10)

The intensity of λ_1 and λ_2 shows the certainty that it's a junction; higher values indicate a junction while low values indicate an irrelevant point.

As noted the tensor voting field depends on two parameters σ and c. The σ parameter defines the size of the attenuation field. The parameter c defines the shape of the attenuation field, smaller values of c result in an attenuation field that is more circular or in other words considers proximity as much as it does continuation when calculating curves. When the value of c is large, the attenuation field is more stick shaped, preferring continuation heavily over proximity when determining curves. The ideal value of c (one which prefers an optimal balance of continuation vs. proximity as defined by Medioni) can also depend directly on σ and is determined by Eq. (7.10) and shown in Fig. 7.19. Figure 7.20 shows the difference between small and large c values and how it affects the voting field's attenuation.

$$c = \frac{-16 \ln(0.1)(\sigma - 1)}{\pi^2} \tag{7.10}$$

Choosing appropriate parameters for the tensor voting framework is of considerable importance for producing good results. The size of the voting field affects how well the tensor voting framework performs. Although relatively insensitive to size parameters, error can be introduced by too large or too small of a voting field. Too large a field can ruin subtle details and features. Too small a field does not adequately remove noise or fill gaps in the raster image. To ensure we have a proper field size and continuation vs. proximity, methods must be developed to approximate an ideal σ and c for each point in the sparse tensor field prior to voting.

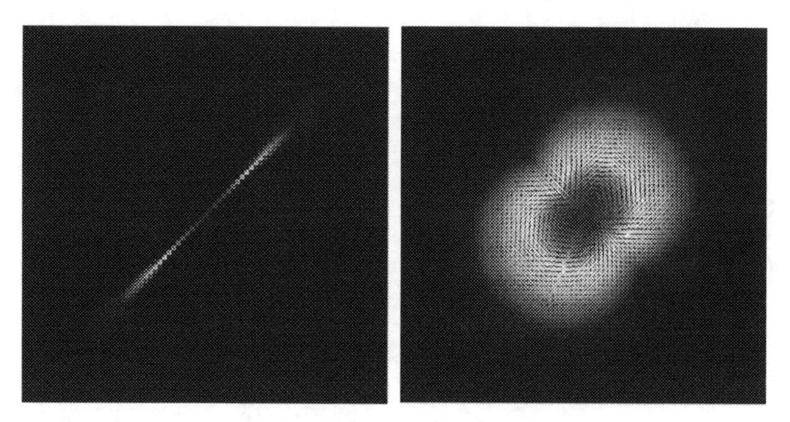

Fig. 7.20 2D view of a voting field's attenuation, *left* $c = 10^7$ vs. *right* $c = 0.1$

7.5.1 Dynamic *c* Values

As shown in Fig. 7.20, we can vary the c values in order to prefer continuation vs. proximity. This can be valuable if we have an insight into what type of feature we are trying to reinforce. By adjusting the c values to make the value higher on long straight roads we can better fill in gaps (even very large gaps). By making the c values small for roads in towns we can better account for the influence of proximal road segments and preserve smaller features.

Dynamically adjusting the c values is done by taking a histogram of the tangent direction in the tensors surrounding a point. The window size of the histogram is based on the value of σ and found by solving for W_{size} in Eq. (7.7). If the mass of the histogram is contained in a small number of bins then c is made higher. If the mass of the histogram is spread out over all the bins and fairly equal, then the c approximates the ideal c value as defined in Eq. (7.10).

7.5.2 Dynamic σ Values

The parameter σ adjusts the size of the attenuation field. Larger attenuation fields reinforce features with more extent while smaller attenuation fields are necessary to preserve features with smaller details. To adjust the σ dynamically, a count of the features within a window is made in order to determine an estimate of the ideal size of the attenuation field that will preserve smaller features and will reinforce extended ones.

If there is a large number of curves or intersections within the window, then the voting field size may be adjusted smaller in order to reinforce fine details and

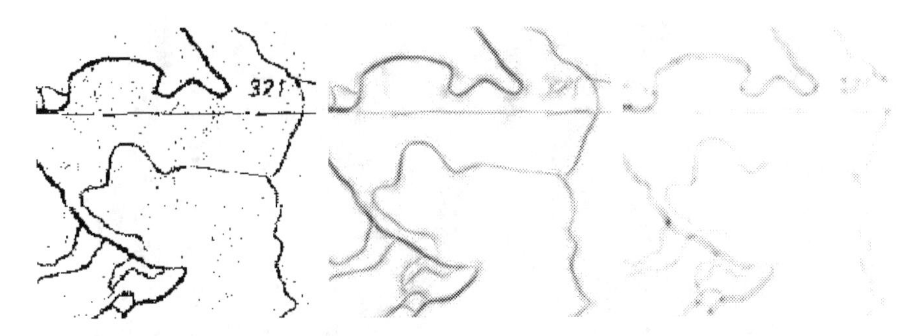

Fig. 7.21 Initial image (*left*), curve map (*middle*), junction map (*right*)

adequately remove noise. However if the ratio of features to the window size is small it can be assumed that there is a limited number of larger features in the window and thus a larger attenuation field may be used.

7.6 Post-processing Techniques

The output of the tensor voting system is a pair of real valued curve and junction maps. From this information the system must derive a binary raster map of the curves (roads) found, then combine this information with the junction map to determine the junctions (road intersections). This is complicated by the fact that within the curve and junction maps, the likelihood of a feature is relative to its surrounding likeliness. For instance, a curve being filled in by the tensor voting system will have a much lower value within the curve map than a curve which exists in the pre-processed image. Therefore, we must take a smart approach to interpreting the data from these systems.

As seen in Fig. 7.21 the junction map and curve map require robust methods to extract information from these raster images. Several different methods are used to extract the features. A process of thresholding and thinning has proved useful, however it misses weak curves in the curve map and isn't very effective on junction maps. The process of using a local maximum algorithm that looks at a window, uses the tensors to determine the normal direction to the curve and then finds the local maximum along the normal has proved useful, but fails at curve pixels closer to junction points. The process of a local maximum is mostly useful for finding junctions. Above all, the most effective algorithm is the process of using connected components and knowledge of the features to find curves and junctions.

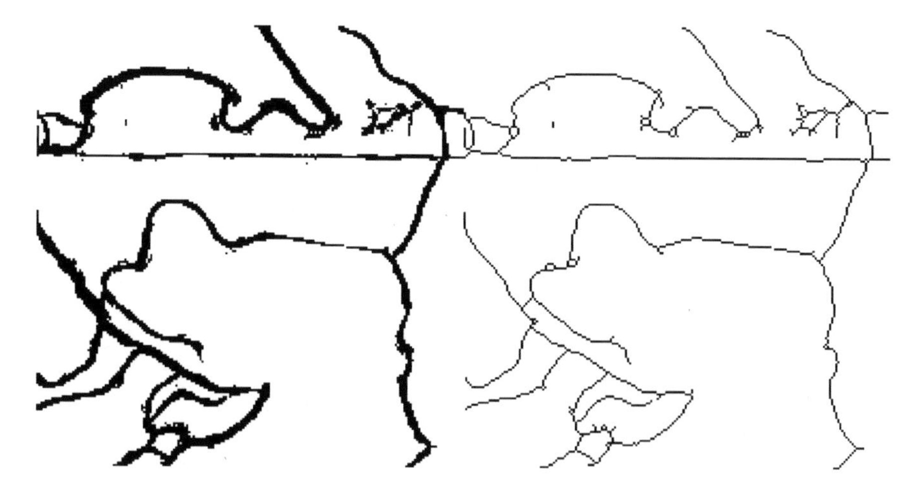

Fig. 7.22 Threshold applied image (*left*), thinned and cleaned image (*right*)

7.6.1 *Global Thresholding and Thinning*

The process of thresholding and thinning as a post processing technique begins by calculating an optimal thresholding value from the histogram of the curve and junction maps. To do this an iterative selection method is used. The method begins by finding the maximum value. The histogram is then split into to parts based on the half the maximum value. The mean of the intensities in the two halves are then calculated, and the average of the two means produces a new threshold. The process is then repeated with the new value until the threshold converges.

The result of the thresholding can be seen in Fig. 7.22 (left). The binary image can then be cleaned and thinned in order to produce a unit width vector feature space (as seen in Fig. 7.22 (right)). Thresholding and thinning still have particular problems making it unsuitable for our needs. Thresholding can remove lines and curves in the feature space which are (relatively) pronounced but in context of the entire image very weak. Thinning (along with skeletonization) can also introduce artifacts as seen in Fig. 7.22 (right); e.g. the thinned image typically has looped nodules and light noise still apparent in the final image.

Even though thresholding and thinning doesn't usually provide a perfect solution, it does produce reasonable features and is very efficient in terms of memory requirements.

7.6.2 *Local Thresholding and Thinning*

The process of local thresholding works similar to global thresholding; however, the method uses a window of 40 by 40 pixel area rather than the entire image.

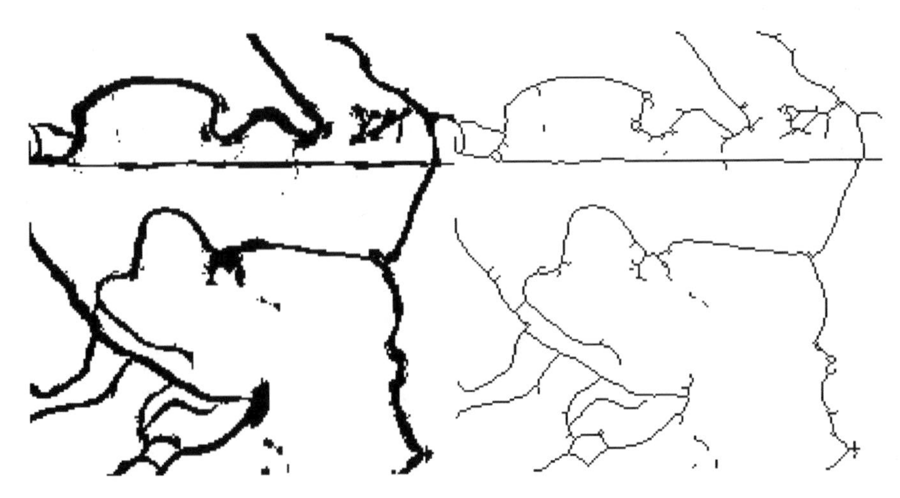

Fig. 7.23 Curve map processed by local thresholding and cleaning. Threshold applied image (*left*), thinned and cleaned image (*right*)

This process helps preserve features where they may be removed by a global thresholding method. The results of the local thresholding can be seen in Fig. 7.23.

Local thresholding has similar problems to global. Artifacts are still apparent in the thinned image and some features are removed creating gaps in the data. Local thresholding also introduces noise to an image that must be resolved.

7.6.3 Local Normal Maximum

Finding the local maximum in the curve map may allow us to figure out the best curve while using the surrounding context of each pixel to determine whether or not the curve passes there. This is important since the likelihood of many reinforced features may not be very high relative to the global curve map, but significantly higher then their neighbors. The process of taking the maximum in the normal direction first identifies the normal axis for each tensor point in the image. Once this is done, the likelihoods are checked along the sides of the current pixel to see if it is in fact a maximum.

The results of this process can be seen in Fig. 7.24. While the process does a fairly good job at identifying curves, it has two major problems. First for pixels close to junctions, good curve pixels may not be the local maximum due to interference from other nearby curves. This causes disconnected curves in the output.

The other problem with this approach is that small differences in the likelihood that are caused by noise get profoundly exaggerated by this process since its goal is to look for relative maxima in likelihood. While this could be solved by thresholding the image (granted less than what would be needed in thresholding the image for thinning) it eliminates smaller roads and less defined features we want to preserve.

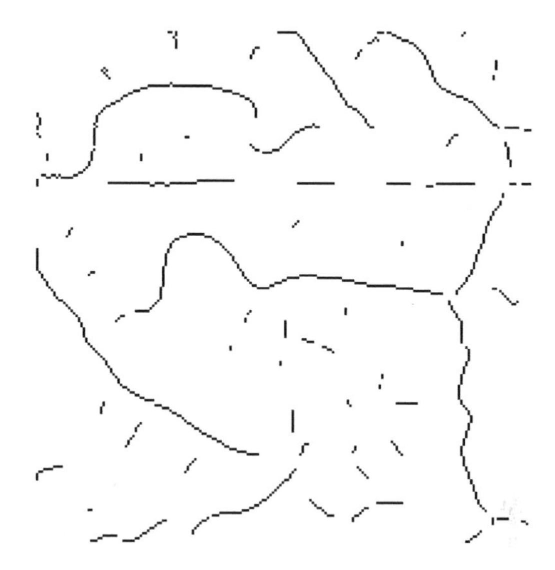

Fig. 7.24 Results of looking for the local maxima over the normal

Due to the drawbacks described, this algorithm it is not effective for our needs. It is effective, however, in finding junctions by looking for local maxima in a local area.

7.6.4 Knowledge-Based Approach

An effective method to find the best paths in the curve map is a connected component approach. The method begins by producing a binary version of the curve map using local thresholding. The resulting raster image is then thinned. While this produces artifacts, noise and raises issues similar to thresholding and thinning described above, the method then uses a connected component and knowledge based algorithm to remove noise and clean the image. This eliminates most of the artifacts in the image.

The thinned image is broken up by its intersections and then processed by running a connected components analysis. Once the image has been broken up into connected components, end points are found as follows. To find end points a 3x3 image window is taken around each defined pixel and run against a look up table of end points. Each segment from the connected component analysis is examined to see if it lies on an endpoint. If it does and the line does not cross a border of the image and it is smaller then 15 pixels (a number derived from the maximum size found in noise produced by the process) then it is removed. The results of this process can be seen in Fig. 7.25. Any segments found in the connected component analysis not connected to the border of the image are also removed. A combination of logical decisions based on the connected components can decide whether certain components should be removed and considered noise.

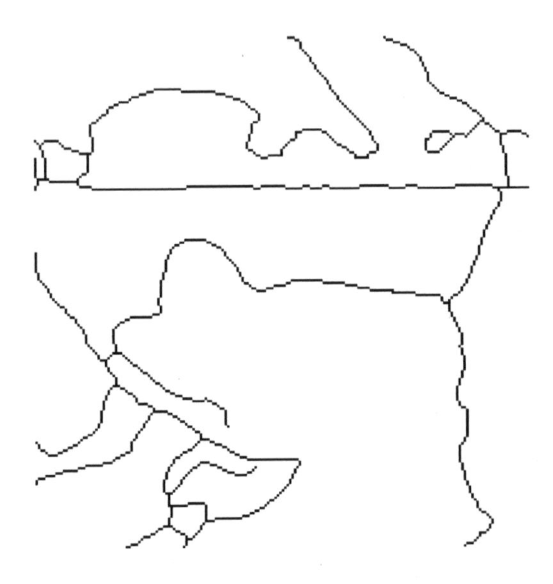

Fig. 7.25 Results from thresholding, thinning and knowledge-based approach using connected component analysis

7.7 Experiments

7.7.1 Method

To determine the performance of the tensor voting, pre-processing and post-processing methods, a framework for evaluating recall and precision is necessary. To evaluate each method a set of one hundred 200x200 pixel image samples are taken from various USGS maps. The maps are then processed through the framework and examined against a ground truth for both the roads and intersections.

Experiments are selected to quantitatively answer questions related to performance. Considerations beyond which method performed best are evaluated, and the goals for these experiments are as follows:

- Determine adequate parameters for pre-processing, post-processing methods and tensor voting.
- Identify weaknesses and strengths of each method.
- Determine the best performing pre-processing method.
- Determine the best performing post-processing method.
- Quantify the contributions of tensor voting.
- Characterize classification distortion of the pre-processing, tensor voting and post-processing methods on perfect inputs.
- Determine the impact of misclassification of text as roads.

The results for each experiment are given with a recall and precision metric for both roads and intersections. Recall and precision are defined by Eq. (7.1). *relevant* is the set of pixels which belong to the class and *retrieved* is the set of pixels which are classified as being in the class. Recall measures the system's capability to find features while precision characterizes whether it was able to find only those features; both are measured on a scale from 0 to 1 where 1 (for both recall and precision) is best. A pixel classified by an algorithm is said to correspond to a feature in the ground truth set if the two are within a specified Euclidean distance (usually 5 pixels).

7.7.2 Data Selection

Data selection is a key aspect of creating accurate and replicable results. To properly select a data set to use in the evaluation, the following constraints are considered:

- The data set must be a large enough to adequately represent features.
- One sub-image of the data must not be biased by the selector.
- One sub-image may not overlap another.
- A sub-image may not be a portion of the map which contains map borders, margins or the legend.

To meet the constraints of data selection a system was built in order to select adequate data sets from two USGS maps (F34086A1.TIF and F34086E1.TIF). The system began by randomly selecting 100 sub-images from the maps, each with are 200x200 pixels. The system was constrained to only select regions which were inside the boundaries of the map so that items such as the legend and meta-data would not be included in the test samples. The system then went through each map in the sample set and checked for duplicates. The same number of images were selected from both test maps.

7.7.3 Ground Truth

Ground truth defines a baseline of what would qualify as a correct answer for both roads and intersections. Once the data selection system finishes the ground truth is generated manually. Each image generated by the data selection process is examined and a binary raster map (e.g., see Fig. 7.26) is made from it by identifying where the roads exist. Similarly the intersections are manually identified and stored in a structured data set. To identify roads the user refers to the legend to make distinctions between geopolitical lines, water, contour lines and roads of all types.

To ensure the ground truth is accurate, the masks are generated twice and the difference between the two is used to identify problems. The ground truth pairs which contained significant differences (above 5%) are re-examined. Neither the original raster map nor the ground truth images were modified so as not to affect the results or bias them for the system. All features defined as roads in the ground truth are marked by a 3x3 pixel line.

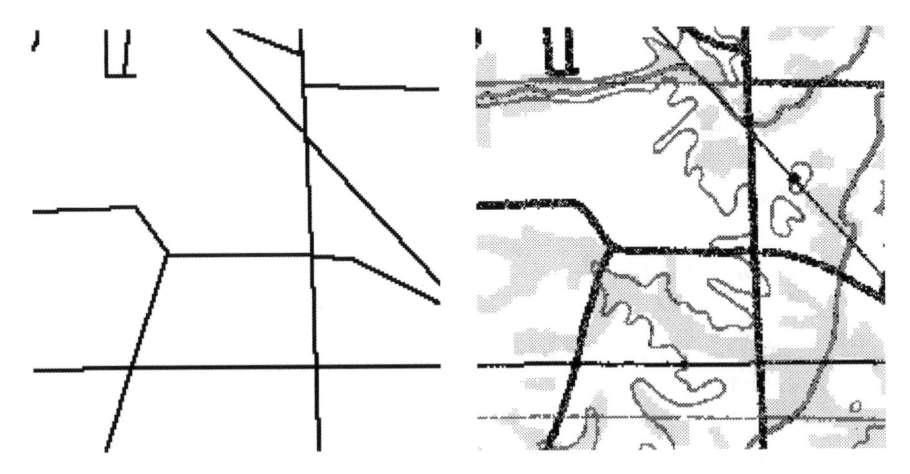

Fig. 7.26 Sample ground truth (*left*) and original image (*right*) from F34086A1 map

7.7.4 Pre-processing

The three pre-processing methods examined are k-Nearest Neighbors, Class Conditional Density Classifier and the Knowledge-based Classifier. The only parameter investigated and varied out of these methods is k in k-Nearest Neighbors. The window size of both k-Nearest Neighbors and Class Conditional Density Classifier was examined in a qualitative manner to determine the best window size (3x3). To determine the best pre-processing method, an appropriate value must be selected for the parameter k. To find the best value for k, k-Nearest Neighbors is run on the ground truth images over a range of acceptable values to determine which value produces the highest recall and precision. The recall and precision measurements are only calculated on the output of k-Nearest Neighbors and no post-processing methods nor tensor voting system is used. Tests run on k from 1 to 20 show little (less than 1%) difference in the recall and precision of the system. The highest value was shown to be at 10 which produced an average road recall of 100%, average road precision of 87%, average intersection recall of 49% and average intersection precision of 11%.

Table 7.6 shows quantitatively that the Knowledge Classifier performed the best (adding together road recall, road precision, intersection recall and intersection precision) and produced similar recall to the other pre-processing techniques but much better precision. The Class Conditional Density Classifier seemed to overclassify items which were more prevalent in the histogram models yet ignore smaller features, the effect of this is shown in Fig. 7.27. While this issue could possibly be resolved by using different histogram models the results for the CCD on various models yielded only small differences in its recall and precision. k-Nearest Neighbors did fairly well but still had more noise than the Knowledge Based Classifier.

Table 7.6 Comparison of pre-processing methods without tensor voting or post-processing method

Method	RR	RP	IR	IP
k-nearest neighbors	100%	87%	49%	11%
Class conditional classifier	100%	86%	46%	9%
Knowledge classifier	99%	92%	51%	17%
Naïve (selecting black)	100%	86%	53%	13%

(*RR*, road recall; *RP*, road precision; *IR*, intersection recall; *IP*, intersection precision)

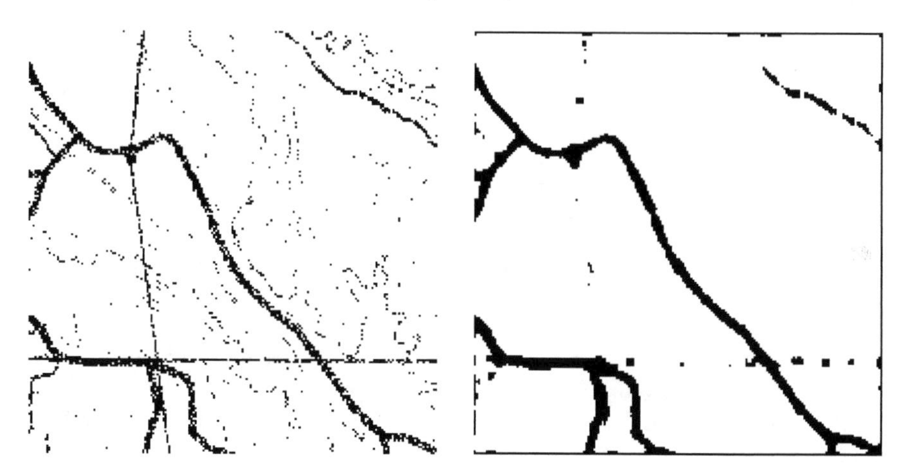

Fig. 7.27 Class conditional density example, original image (*left*) with roads identified (*right*)

The Knowledge-based Classifier did better than any other method; however, it had difficulties finding smaller dirt and utility roads in the image and introduced enough noise to lower its precision. The naïve approach of selecting all black did considerably well coming performing slightly better than *k*-Nearest Neighbors and the Class Conditional Density Classifier. This was mainly because most road textures end up being primarily black, in addition to this all pre-processing method results go through an open and close morphological operation which tends to fill the majority of the minor gaps and filter out noise in the naïve method.

7.7.5 Tensor Voting

The Tensor Voting system is examined to find its contributions to the overall recall and precision of the framework. To do this an appropriate value for the parameter σ must be found. To find the best value the Tensor Voting system is run on the ground truth images with a range of σ's. Because the Tensor Voting system requires a pre-processing method and post-processing method in order to function, the naïve pre-processing method (selecting all black) and the Knowledge-based Approach post-processing method are used. These were selected because they do not require parameters and do not need to be tuned for best performance; therefore,

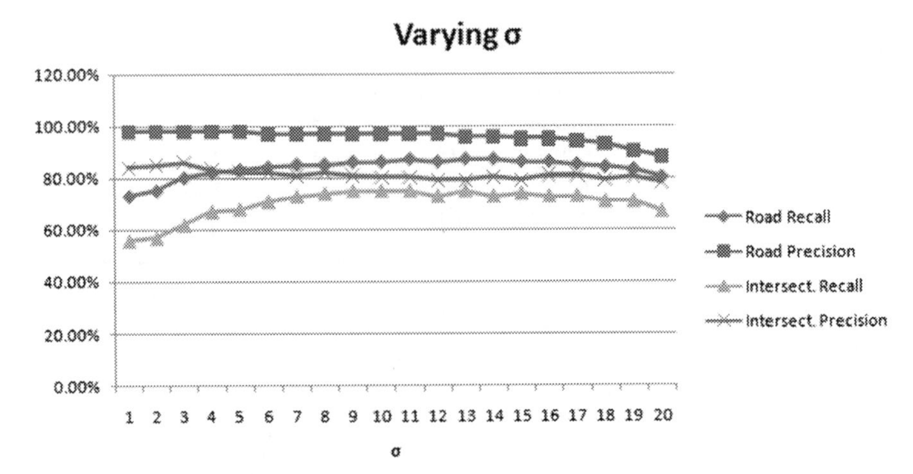

Fig. 7.28 Affect of varying the sigma parameter of the tensor voting system on recall and precision

we can accurately determine the affect of various σ's on the Tensor Voting system. Figure 7.28 shows the affect of varying the σ parameter on recall and precision.

The results from this test show that the best value for σ is between 10 and 16 with little difference in the performance between these ranges. The value 13 was used as the parameter for σ for all further tests.

To determine the Tensor Voting system's contribution to the overall performance, the framework is run on the ground truth tests with tensor voting and without. The framework is run for all combinations of pre-processing and post-processing methods to determine the contribution of Tensor Voting for each combination. Table 7.7 shows the results of recall and precision for all combinations of pre-processing and post processing methods with Tensor Voting. Table 7.8 shows the results of recall and precision for all combinations of pre-processing and post processing methods without Tensor Voting.

From the data in Tables 7.7 and 7.8 one can determine that certain pre-processing and post-processing method actually perform better without tensor voting. Specifically the global and local thresholds actually perform better simply because they receive a binary output from the pre-processing method, which is very easy to threshold. When used with Tensor Voting the global and local thresholds must threshold a gray-scale image to a binary image, therefore it's somewhat expected that the results would be better for the local thresholding and global thresholding without tensor voting. The same can be said for the naïve method.

However, the Knowledge-based Approach benefitted from using the Tensor Voting system, simply because of its effectiveness in using the output of the Tensor Voting system to accurately identify features. On average the road recall for the Knowledge-based Approach increased by 10%, but the road precision dropped by 2%. The largest benefit is the intersection recall increased by 22%, and the intersection precision increased by 20%.

Table 7.7 Comparison of pre-processing (column) and post-processing (row) methods with tensor voting

	k-nearest neighbors	Class conditional classifier
Knowledge-based	92% / 95% / 82% / 80%	81% / 98% / 66% / 90%
Local thresh	99% / 77% / 85% / 26%	98% / 77% / 85% / 26%
Global thresh	98% / 75% / 82% / 21%	98% / 75% / 82% / 21%
Local max	77% / 40% / 78% / 75%	73% / 31% / 71% / 73%
Naïve	100% / 27% / 52% / 5%	100% / 28% / 52% / 5%
	Knowledge classifier	Naïve (selecting black)
Knowledge-based	77% / 98% / 65% / 86%	86% / 96% / 73% / 80%
Local thresh	98% / 87% / 84% / 29%	99% / 79% / 86% / 30%
Global thresh	97% / 84% / 83% / 22%	99% / 78% / 83% / 25%
Local max	77% / 43% / 77% / 73%	71% / 33% / 70% / 71%
Naïve	100% / 31% / 52% / 6%	100% / 26% / 52% / 4%

(Each cell contains the percentage for (Road Recall / Road Precision / Intersection Recall / Intersection Precision)

Table 7.8 Comparison of pre-processing (column) and post-processing (row) methods without tensor voting

	k-Nearest neighbors	Class conditional classifier
Knowledge-based	82% / 97% / 60% / 60%	80% / 98% / 57% / 83%
Local thresh	99% / 77% / 88% / 26%	98% / 77% / 89% / 33%
Global thresh	99% / 77% / 88% / 26%	98% / 77% / 89% / 33%
Local max	NA / NA / NA / NA	NA / NA / NA / NA
Naïve	100% / 87% / 49% / 11%	100% / 86% / 46% / 9%
	Knowledge classifier	Naïve (selecting black)
Knowledge-based	77% / 98% / 50% / 86%	79% / 98% / 57% / 85%
Local thresh	97% / 92% / 87% / 38%	100% / 81% / 70% / 14%
Global thresh	97% / 92% / 87% / 38%	100% / 81% / 70% / 14%
Local max	NA / NA / NA / NA	NA / NA / NA / NA
Naïve	99% / 92% / 51% / 17%	100% / 86% / 53% / 13%

(Each cell contains the percentage for (Road Recall / Road Precision / Intersection Recall / Intersection Precision)

7.7.6 Post-processing

The four post-processing methods examined are the Knowledge Based Approach, Local Thresholding, Global Thresholding and Local Normal Maxima. The only parameter which can be varied out of these methods is the window size of the Local Thresholding. To determine the best post-processing method, an appropriate value must be selected for the window size. To find the best value for it, the Local Thresholding method is run on the ground truth images over a range of acceptable values to determine which value produces the highest recall and precision. Since the post-processing techniques require some sort of pre-processing technique and

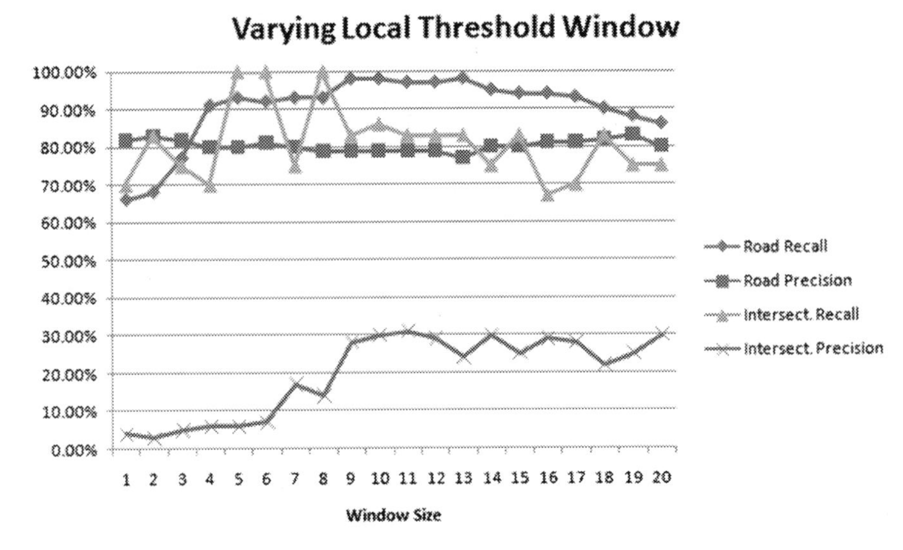

Fig. 7.29 Affect of varying the window size of the local thresholding method on recall and precision

Table 7.9 Comparison of post-processing methods with tensor voting and a naïve (selecting all black) pre-processing method

Method	RR	RP	IR	IP
Knowledge-based	86%	96%	73%	80%
Local thresh	99%	79%	86%	30%
Global thresh	99%	78%	83%	25%
Local max	71%	33%	70%	71%
Naïve	100%	26%	52%	4%

(*RR*, road recall; *RP*, road precision; *IR*, intersection recall; *IP*, intersection precision)

require tensor voting, both are used; however, the naïve (selecting all black) is used for the pre-processing method. The tensor voting uses the best value for its σ parameter as previously determined. The results of this test can be seen in Fig. 7.29. The best value for the window size is determined to be between 10 and 14.

Qualitatively the Knowledge-based System performed the best, and Table 7.9 shows quantitatively that the Knowledge-based Approach performed the best. Local thresholding along with global thresholding can produce higher numbers in recall but lower in precision. This is because liberal thresholding lowers the precision, but conservative thresholding lowers recall. The local maximum had a very low recall and precision; this is expected due to the inability of the local maximum to find all the features.

Table 7.10 Comparison of pre-processing (column) and post-processing (row) methods with tensor voting

	k-Nearest neighbors	Class conditional classifier
Knowledge-based	92 % / 95% / 82% / 80%	81% / 98% / 66% / 90%
Local thresh	99% / 77% / 85% / 26%	98% / 77% / 85% / 26%
Global thresh	98% / 75% / 82% / 21%	98% / 75% / 82% / 21%
Local max	77% / 40% / 78% / 75%	73% / 31% / 71% / 73%
Naïve	100% / 27% / 52% / 5%	100% / 28% / 52% / 5%
	Knowledge classifier	Naïve (selecting black)
Knowledge-based	77% / 98% / 65% / 86%	86% / 96% / 73% / 80%
Local thresh	98% / 87% / 84% / 29%	99% / 79% / 86% / 30%
Global thresh	97% / 84% / 83% / 22%	99% / 78% / 83% / 25%
Local max	77% / 43% / 77% / 73%	71% / 33% / 70% / 71%
Naïve	100% / 31% / 52% / 6%	100% / 26% / 52% / 4%

(Each cell contains the percentage for (road recall / road precision / intersection recall / intersection precision)

7.7.7 *Best Combination*

To determine the best combination of pre-processing and post-processing methods they are combined and run over the ground truth set to see which combination produces the best results. The four pre-processing methods: *k*-Nearest Neighbors, Class Conditional Density Classifier, Knowledge Based Classifier and the naïve method (selecting all black) are each run against the five post-processing methods: Knowledge Based Approach, Local Thresholding, Global Thresholding, Local Maxima and the naïve method (selecting everything above 0 as a curve). This yields 80 results, (20 combinations and 4 results for each combination).

Table 7.10 shows the results of the test. *k*-Nearest Neighbors as a pre-processing method and Knowledge-based Approach as a post-processing method performed the best with a road recall of 92%, road precision of 95%, intersection recall of 82% and intersection precision of 80%. For a 95% confidence interval the road recall was [92.47%,94.75%] $\pm$ 0.14%, the road precision was [94.13%,96.33%] $\pm$ 0.10%, the intersection recall was [78.91%,85.51%] $\pm$ 3.29% and the intersection precision was [76.31%,82.99%] $\pm$ 2.89%.

7.7.8 *Perfect Data*

To determine the affects of post-processing method and the tensor voting system on perfect data the framework was run on the set of ground truth as input. Table 7.11 demonstrates the results of each method run on the ground truth, against the ground truth. The pre-processing methods were not compared or used in the perfect data analysis since the ground truth images are already labeled with the roads, since

Table 7.11 Comparison of post-processing techniques with tensor voting on perfect data

Method	RR	RP	IR	IP
Knowledge-based	94%	100%	80%	86%
Local Thresh	100%	98%	86%	46%
Global Thresh	100%	96%	83%	36%
Local Max	80%	77%	75%	72%
Naïve	100%	37%	52%	8%

(*RR*, road recall; *RP*, road precision; *IR*, intersection recall; *IP*, intersection precision)

pre-processing methods use histogram models to determine initial estimates of labels they are not used.

The results show better recall and precision than any other method or combination of methods that are examined here. The precision of some methods are considerably low due to minor road noise which is introduced by the post-processing methods that created significant intersection precision problems. The naïve, global and local thresholding have this problem; however, the knowledge based approach did considerably better at not introducing any noise.

7.7.9 No Text

Extraction of roads is complicated by features in the raster image which have the same texture as roads. Text identifying meta-data in the map has similar (if not the same) texture as roads, and to extract these a method which relies on structure rather than texture is necessary. To compute the effect of text on the ground truth, the text was manually removed from each sample image and re-run on the ground truth. Table 7.12 shows the results (Recall and Precision) of the samples once text has been manually removed.

Overall, removing the text from the map data increased the recall by 1% for roads and 2.5% for intersections, the precision wasn't significantly affected (less than 1% difference). The best combination was still k-Nearest Neighbors and Knowledge-based Approach for the pre-processing and post-processing systems. The road recall was 94%, the road precision was 95%, the intersection recall was 83% and the intersection precision was 80%.

7.7.10 Comments

The quantitative examination of pre-processing and post-processing systems showed k-nearest neighbors and the knowledge-based approach to be the best combination

Table 7.12 Comparison of pre-processing and post-processing methods run on ground truth images without text

	k-Nearest neighbors	Class conditional classifier
Graph search	94% / 95% / 83% / 80%	83% / 77% / 71% / 80%
Local thresh	99% / 76% / 85% / 26%	99 % / 76% / 85% / 26%
Global thresh	99% / 75% / 81% / 21%	98% / 75% / 83% / 22%
Local max	77% / 40% / 78% / 75%	73% / 31% / 71% / 73%
Naïve	100% / 26% / 52% / 5%	100% / 27% / 52% / 5%
	Knowledge classifier	Naïve (selecting black)
Graph search	80% / 85% / 68% / 80%	89% / 96% / 78% / 84%
Local thresh	98% / 87% / 84% / 29%	99% / 79% / 87% / 31%
Global thresh	98% / 84% / 83% / 23%	98% / 77% / 84% / 25%
Local max	77% / 42% / 78% / 73%	72% / 33% / 71% / 71%
Naïve	100% / 30% / 52% / 5%	100% / 25% / 52% / 5%

(road recall / road precision / Int. recall / Int. precision)

found. The text in the raster maps had a significant impact on precision especially on intersections but did not change the results of recall dramatically. Interestingly, the Knowledge-based Classifier as a pre-processing method was found to be the best pre-processing method when no post-processing method or tensor voting system was used. However, when run in combination with others it didn't perform as well as other pre-processing methods. The reasons behind this are somewhat complicated, the Knowledge-based Classifier produced a larger amount of sparse noise in the images than any other method which caused the tensor voting system and some post-processing systems to inaccurately identify curves. While both the Class Condition Classifier and *k*-Nearest Neighbors both were almost as effective but performed worse in the independent pre-processing analysis, neither introduced the same type of noise (it was more dense rather then sparse) which affects the tensor voting system in a way which instead of removing the noise, actually exaggerates it.

Dynamically adjusting the σ and c parameters for the tensor voting produced insignificant results. The σ was varied from 4 to 20 by a dynamic σ system described previously, the results of dynamically adjusting the σ however were somewhat irrelevant to the overall performance only affecting the average recall and precision by $\pm$ 1% . Dynamically adjusting the c value for the tensor voting actually produced worse performance. The average decline in performance over all methods was 18.3%. This was due to slight variations in the tangent direction for each tensor in the field. If a slight variation occurs while the c prefers a stick vote it tends to produce misaligned roads and lines between points in the final curve map produced.

Fig. 7.30 Sample of intersecting roads that are too close

7.8 Conclusions and Future Work

A method of *k*-nearest neighbors, tensor voting and the knowledge based approach produced the best results for both roads and intersections. *k*-nearest neighbors showed a unique ability to overcome noise, variances in texture for features and to produce smooth results for the tensor voting system. The tensor voting system showed significant capabilities for filling gaps, removing noise and finding accurate junctions in the output raster image. The knowledge based approach was most effective at producing unit width binary raster map from the output curve of the tensor voting system.

The runtime performance of the system was decent, the real time to run the tensor voting on a 200x200 area was 22 seconds on average on a Pentium Dual Core 2.9ghz processor running Matlab. The total runtime of the system was 60.1 seconds for a 200x200 area. It should be mentioned that the implementation of this was not optimized and significant runtime performances could be made through different languages and optimizations.

The largest problems with extracting features was preserving features close together within the raster map. Nearly every method investigated was unable to properly segment features when they became closely grouped with the exception of the knowledge-based classifier which qualitatively did the best job at this. Another significant problem was roads which intersect and are at a small angle to each other or have fairly thick lines that must be thinned. Figure 7.30 shows a road in the lower right hand corner which intersects with another at a small angle and has thick overlapping lines. When this occurs instead of forming one intersection the tensor voting system curves the two roads together prematurely and produces two intersections. The output of the tensor voting system is shown in Fig. 7.31 which demonstrates how the intersection is broken in two.

Text is also a significant challenge to overcome in the raster maps. The text can appear in any orientation, can have different font types, seems to be bolder in some parts then others, overlaps existing features, has the same texture as features

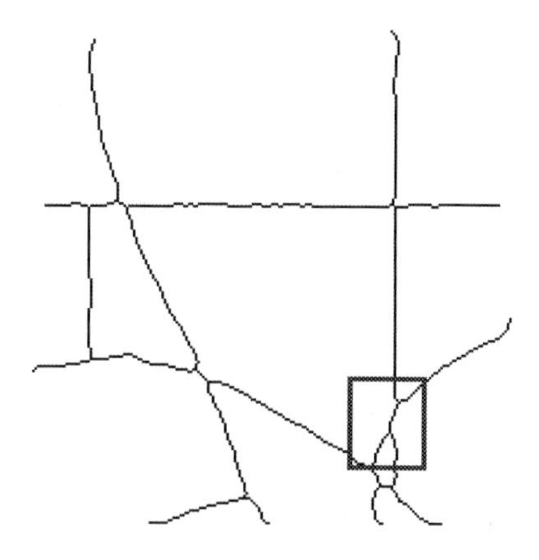

Fig. 7.31 Sample of the output from tensor voting and post-processing method for intersecting roads that are too close

in the raster and has various sizes. In order to over come this, either a method using structural analysis is necessary, or a test recognition and removal step should precede road analysis.

Chapter 8
Other Semantic Feature Segmentation

The automatic classification of semantic classes[1] (background, vegetation, roads, water, political boundaries, iso-contours) in raster map images still poses significant challenges. We describe and compare the results of three unsupervised classification algorithms: (1) k-means, (2) graph theoretic (GT), and (3) expectation maximization (EM). These are applied to USGS raster map images, and performance is measured in terms of the recall and precision as well as the cluster quality on a set of map images for which the ground truth is available. Across the six classes studied here, k-means achieves good clusters and an average of 78% recall and 70% precision; GT clustering achieves good clusters and 83% recall with 74% precision. Finally, EM forms very good clusters and has an average 86% recall and 71% precision.

8.1 Introduction

Digital maps contain a wealth of information which can be used for a variety of applications, including the analysis of cultural features, topographical terrain shape, land use classes, transportation networks, or maps can be registered (conflated) with aerial images in order to localize and identify photo imagery structures. Unfortunately, raster map images are typically encoded in such a way that semantic features are difficult to extract due to noise, error or overlapping features. Semantic features of interest include roads, road intersections, water regions, vegetation, political boundaries, and iso-elevation contours. This is still a difficult problem, although various techniques have been proposed in the past [6, 95, 132]. We have worked on road segmentation and road intersection detection [45, 72].

[1]This chapter is a modified version of "Automatic Segmentation of Semantic Classes in Raster Map Image," [46] contributed by Thomas C. Henderson, Trevor Linton, Sergey Potupchik and Andrei Ostanin.

DOI 10.1007/978-1-4419-8167-7_8,

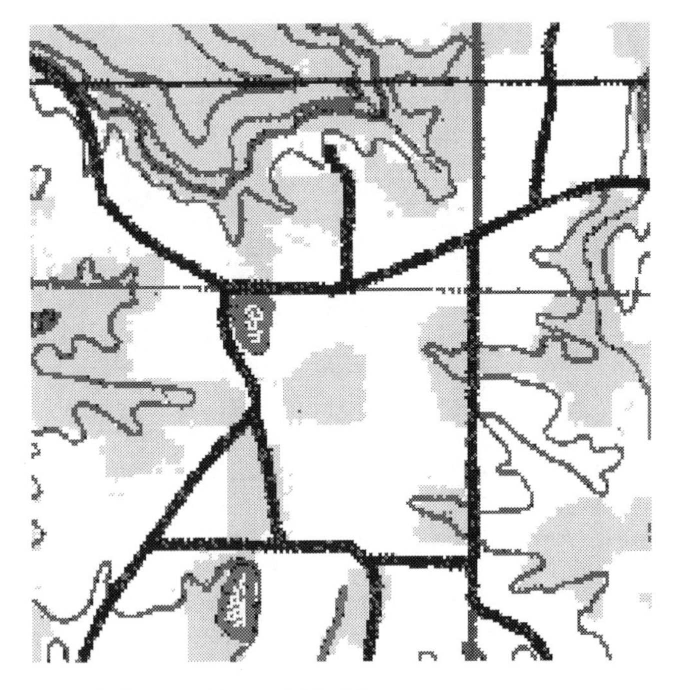

Fig. 8.1 Example USGS map sub-image (200x200)

Our goal is to achieve a semantic segmentation of an arbitrary raster map image through the use of unsupervised classification algorithms. An example USGS map sub-image is shown in Fig. 8.1. We are interested in six basic classes:

- Background
- Vegetation
- Roads
- Water
- Political Lines
- Iso-contours

Figure 8.2 shows the ground truth for these classes for the map in Fig. 8.1.

8.2 Method

The ground truth was determined using a knowledge-based analysis of a set of sub-images (200x200 pixels) taken from ten USGS map images. These maps use six colors (*black*, *white*, *blue*, *red*, *brown*, *green*), and are given as indexed images (i.e., the colors have indexes $0, 1, 2, 3, 4, 5$). The classification analysis process is shown in Fig. 8.3. The index histogram is based on a $w \times w$ window at each pixel.

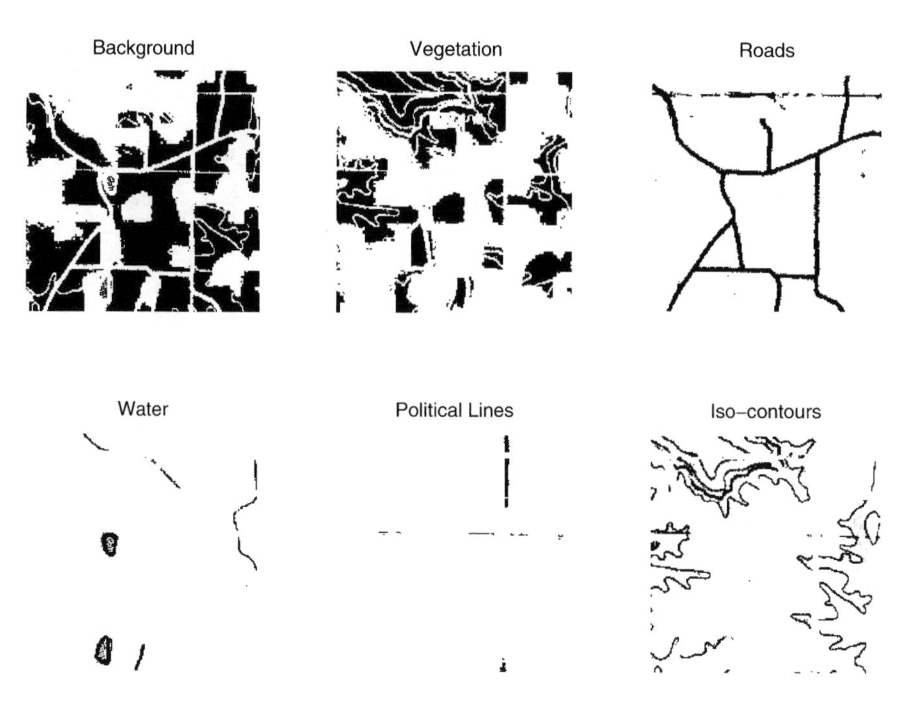

Fig. 8.2 Ground truth classes from example image

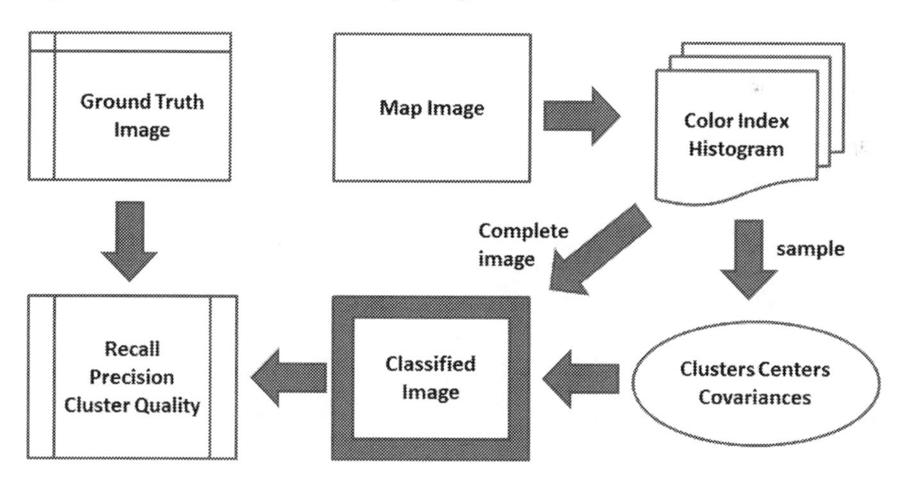

Fig. 8.3 Segmentation analysis process

The cluster centers are the representative histogram for a class, and the covariance matrix gives the variation between the colors for that class. These models are found by using a subset of n samples from the index histogram image. The number of classes may be pre-defined (as with k-means) or determined automatically by the method (e.g., GT). Thus, the parameters of study across the three algorithms are w, the histogram window size, n, the number of samples used to construct the model, and k, the number of classes sought.

The quality measures for the class models are defined in terms of:

- the cluster inter-center distances where, in general, a greater value is better, and
- the distances of points in the cluster from the center where a smaller value is better.

As for the quality of the classification result, recall and precision are defined as:

$$recall = \frac{|relevant \cap retrieved|}{|relevant|}$$

$$precision = \frac{|relevant \cap retrieved|}{|retrieved|}$$

where *relevant* is the set of ground truth pixels in a class and *retrieved* is the set of pixels segmented into that class by the algorithm. The general layout of the classification process is:

```
Algorithm:  Classification Test Process

for each test image
  for each w in Window_sizes
    for each n in Sample_sizes
      for each k in Number_of_classes
        Obtain class centers (means and covariances)
        Compute class quality
        Compute recall and precision
      end
    end
  end
end

Compute statistics over all test images
```

The algorithms under study include k-means, GT and EM. k-means initially selects k random centers, then alternates between assigning points (i.e., histogram vectors) to the nearest center and calculating the centers as the means of the points in the cluster. The graph theoretic method forms an affinity measure between all sample points (e.g., $exp^{-|p_1-p_2|}$), then obtains the eigenvalues and eigenvectors of that matrix; finally, the eigenvectors serve to classify pixels in each class. The EM algorithm alternates between the expectation calculation step and the maximization step to determine the set of classes. See [37] for more details on these three methods.

The centers and covariances are found for each classification algorithm by computing the mean of the sample points segmented into a class, and the covariance of those points. Although these are produced directly by k-means and EM, this is done after the fact for GT based on the set of points in the sample.

The map image is classified by simply labeling each pixel according to the closest center to the pixel's index color histogram. Note that the method cannot know which, if any, of its discovered classes correspond to ground truth classes. Therefore, we determine the recall and precision by mapping each discovered class to the nearest (Euclidean distance) ground truth class mean histogram vector.

8.3 Data

Here we give the results of the Algorithm Classification Test on the three algorithms. The possible values for the parameters were:

- k-means:

$$w \in \{1,3,5\}$$

$$n \in \{1000, 2000, 3000\}$$

$$k \in \{6, 8, 10\}$$

- Expectation Maximization (EM)

$$w \in \{1,3,5\}$$

$$n \in \{1000, 2000, 3000\}$$

$$k \in \{6, 8, 10\}$$

- Graph Theoretic

$$w \in \{1,3,5\}$$

$$n \in \{25, 50, 75\}$$

$$s \in \{0.1, 10, 20\}$$

These w values correspond to a single pixel ($w = 1$) up to a window that almost always includes background with any linear feature. The values of n range from about 25% of linear features in an average 200×200 sub-image, up to the full number of linear features in a typical sub-image. Of course, there is no guarantee

Table 8.1 *k*-means ranked parameter combinations (first three values per row), followed by average recall (over all classes all images), average precision, and sum of average recall and precision

w	*n*	*k*	Mean recall	Mean precision	Recall + precision
1	2,000	8	0.78	0.70	1.48
1	3,000	10	0.78	0.70	1.48
1	2,000	10	0.77	0.70	1.47
1	1,000	10	0.78	0.69	1.47
1	3,000	6	0.79	0.67	1.46
1	1,000	6	0.80	0.66	1.46
1	3,000	8	0.78	0.68	1.46
1	2,000	6	0.80	0.66	1.46
1	1,000	8	0.77	0.68	1.45
3	1,000	10	0.59	0.54	1.13
3	2,000	10	0.58	0.53	1.12
3	2,000	8	0.59	0.52	1.11
3	3,000	10	0.58	0.53	1.11
3	1,000	8	0.59	0.52	1.11
3	3,000	6	0.60	0.49	1.09
3	1,000	6	0.60	0.49	1.09
3	2,000	6	0.60	0.49	1.09
3	3,000	8	0.58	0.51	1.09
5	2,000	10	0.49	0.45	0.93
5	3,000	6	0.52	0.42	0.93
5	1,000	10	0.49	0.45	0.93
5	2,000	8	0.49	0.44	0.93
5	3,000	10	0.48	0.44	0.93
5	1,000	6	0.50	0.42	0.92
5	1,000	8	0.49	0.43	0.92
5	2,000	6	0.50	0.41	0.91
5	3,000	8	0.48	0.42	0.91

that pixels in a linear feature will be selected as samples. The number of classes of interest is six; however, not all classes may be present in a sub-image; moreover, pixels at the boundary of two classes actually represent a different class (e.g., vegetation-water boundary). Finally, the s value is a distance scaling measure in the graph theoretic method which controls the scale of the affinity.

There are 27 combinations of w, n, and k/s values. Tables 8.1–8.3 give the parameters of the top performing combinations and the recall and precision values averaged over all classes and all images. Figure 8.1 shows an example raster map image of size 200×200, while Fig. 8.2 shows the ground truth for this image. Figure 8.4 shows the classes found by k-means; Fig. 8.5 shows the graph theoretic classes, and Fig. 8.6 shows the EM classes.

Table 8.2 Graph theoretic ranked parameter combinations (first three values per row), followed by average recall (over all classes all images), average precision, and sum of average recall and precision

w	n	s	Mean recall	Mean precision	Recall + precision
1	50.0000	0.1000	0.8282	0.7414	1.5697
1	25.0000	0.1000	0.8282	0.7414	1.5696
1	75.0000	0.1000	0.8282	0.7414	1.5696
3	25.0000	0.1000	0.8203	0.6895	1.5098
3	50.0000	0.1000	0.7786	0.6798	1.4585
3	75.0000	0.1000	0.7511	0.6765	1.4276
1	75.0000	20.0000	1.0000	0.2662	1.2662
5	25.0000	0.1000	0.6680	0.5798	1.2478
1	75.0000	10.0000	1.0000	0.2330	1.2330
5	25.0000	10.0000	0.6439	0.5890	1.2329
5	75.0000	10.0000	0.6375	0.5921	1.2295
1	50.0000	20.0000	1.0000	0.2228	1.2228
5	50.0000	10.0000	0.6343	0.5821	1.2164
1	50.0000	10.0000	1.0000	0.1998	1.1998
3	75.0000	20.0000	1.0000	0.1908	1.1908
1	25.0000	10.0000	1.0000	0.1897	1.1897
1	25.0000	20.0000	1.0000	0.1803	1.1803
3	50.0000	20.0000	1.0000	0.1793	1.1793
3	25.0000	20.0000	1.0000	0.1788	1.1788
5	50.0000	0.1000	0.6006	0.5613	1.1619
5	50.0000	20.0000	0.5122	0.6280	1.1402
5	75.0000	20.0000	0.4758	0.6354	1.1112
3	75.0000	10.0000	0.4547	0.6468	1.1016
5	75.0000	0.1000	0.5483	0.5504	1.0987
3	50.0000	10.0000	0.4402	0.6480	1.0881
3	25.0000	10.0000	0.4144	0.6662	1.0807
5	25.0000	20.0000	0.4384	0.6330	1.0713

8.3.1 Test Images

Figure 8.7 shows the ten test images used in the study.

8.4 Future Directions from this Work

The results show that the three clustering methods perform well for unsupervised raster map image classification. Moreover, the optimal parameters all have the window size set to 1×1 (a single pixel). However, the models developed do a little better than simply classifying each pixel based on its color which achieves recall of 80% and precision of 72% for a sum of 1.52; this is worse than graph theoretical and EM, but better than k means.

Table 8.3 Expectation maximization (EM) ranked parameter combinations (first three values per row), followed by average recall (over all classes all images), average precision, and sum of average recall and precision

w	n	k	Mean recall	Mean precision	Recall + precision
1	3,000	8	0.8644	0.7054	1.5697
1	2,000	8	0.8681	0.6850	1.5532
1	3,000	10	0.8515	0.7011	1.5526
1	1,000	10	0.8482	0.7023	1.5505
1	2,000	10	0.8503	0.7001	1.5504
1	1,000	8	0.8549	0.6897	1.5446
1	1,000	6	0.8710	0.6719	1.5429
1	2,000	6	0.8784	0.6637	1.5421
1	3,000	6	0.8764	0.6628	1.5392
3	2,000	8	0.9550	0.2114	1.1664
3	3,000	10	0.9519	0.2123	1.1642
3	2,000	10	0.9518	0.2120	1.1638
3	1,000	8	0.9540	0.2093	1.1633
3	3,000	8	0.9568	0.2063	1.1630
3	1,000	6	0.9577	0.2034	1.1611
3	1,000	10	0.9509	0.2081	1.1589
3	3,000	6	0.9599	0.1986	1.1584
3	2,000	6	0.9599	0.1948	1.1547
5	1,000	6	1.0000	0.0171	1.0171
5	1,000	8	1.0000	0.0171	1.0171
5	2,000	6	1.0000	0.0171	1.0171
5	3,000	6	1.0000	0.0171	1.0171
5	3,000	8	1.0000	0.0171	1.0171
5	2,000	8	0.9716	0.0176	0.9892
5	3,000	10	0.9541	0.0187	0.9728
5	1,000	10	0.9455	0.0186	0.9641
5	2,000	10	0.9408	0.0188	0.9596

The fact that a small number of samples can be used is also good; the graph theoretic method must calculate the eigenvalues of an $n \times n$ affinity matrix, and thus, the lower n, the better.

Of course, these are relatively simple raster map images with only six colors. It is necessary to study these methods on map images with more colors. This will increase the length of the histogram vectors unless some form of color clustering is performed first to reduce the number of color classes. This may require conversion to a color representation with a reasonable distance metric between colors (i.e., where various types of blue are close in the metric space).

Another issue worthy of study is a more informed method to select samples. It may be worthwhile to ensure that samples represent the variety of classes in the image (as opposed to the standard sampling goal of proportional representation of the sampled population). It may be possible to use edge detection to distinguish class boundary pixels or texture parameters to determine classes expressed as textures.

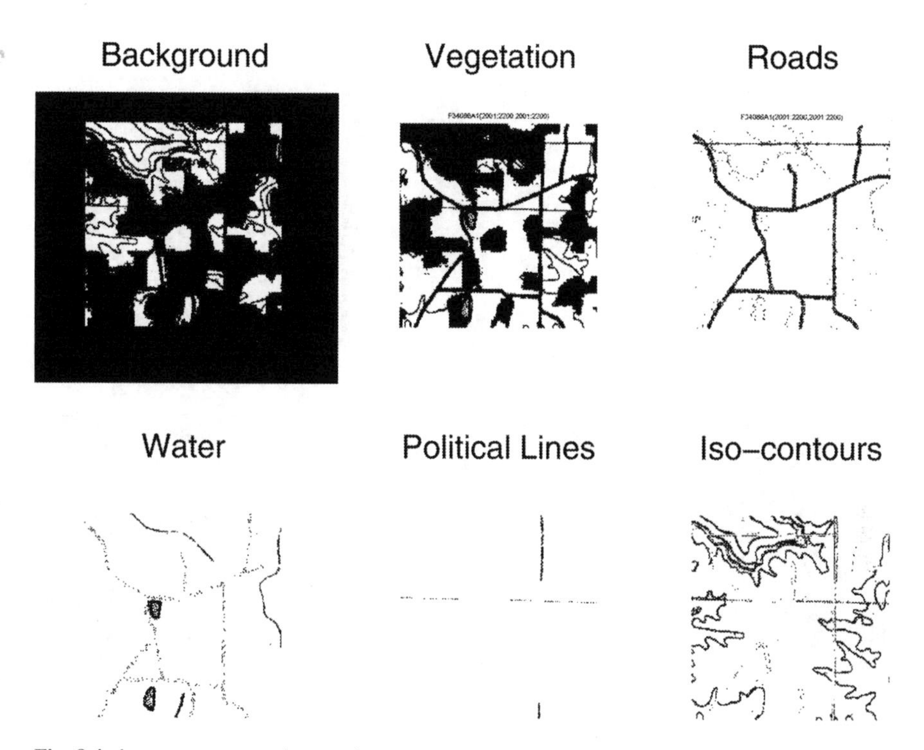

Fig. 8.4 *k*-means segmentation results

Of course, edge and texture information may be included in the feature vector (in addition to the color histogram).

Finally, all these classification methods have a variety of possibilities in algorithm implementation. Initialization methods, re-starting empty classes, thresholds, and distance measures all offer a number of options which should be studied.

8.5 Other Semantic Segmentation Approaches

8.5.1 Water Segmentation

Linear waterways (e.g., rivers, creeks, canals, etc.) are mainly comprised of **BLUE** pixels (index equals 2 in USGS raster images). However, many other linear features overlay water in the map (e.g., iso-contours). In addition, names of waterways, lakes, etc. are also **BLUE**, and must be distinguished from waterways proper. (Text is handled separately.) Finally, **BLUE** pixels may be found in other features, like primary highways, various textures, and even as noise pixels.

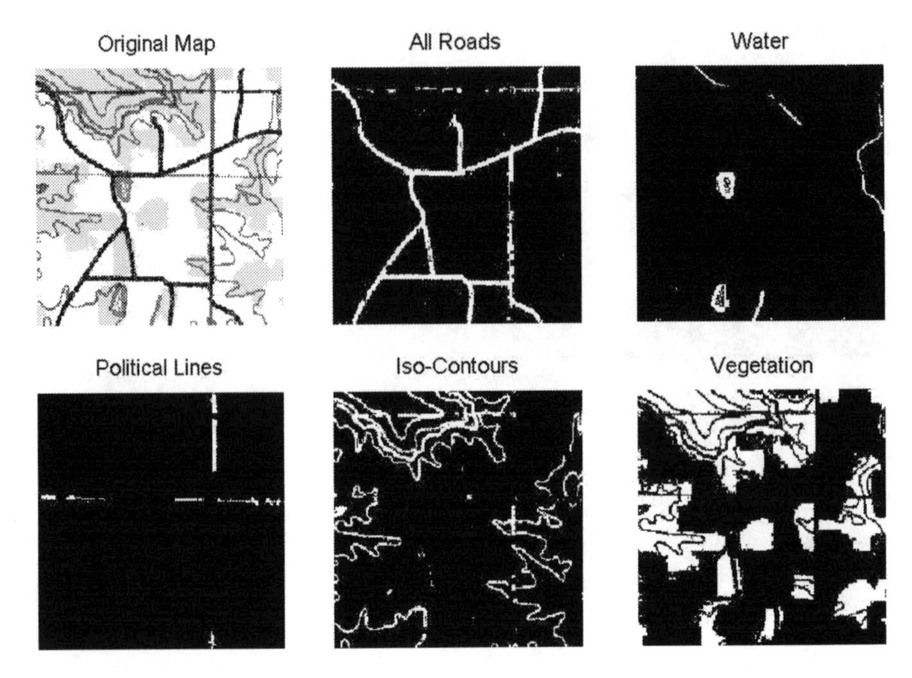

Fig. 8.5 GT segmentation results

Water Segmentation Algorithms

The segmentation of linear water features is performed as follows:

- If the **BLUE** or **WHITE** pixels constitute the majority in a (5x5) window, and the pixel is **BLUE**, then the pixel is classified as *water*.
- A median filter is run over the *water* image.
- For every **BLUE** pixel in the original map image, if there are *water* pixels in its 5x5 neighborhood in the *water* image, then it, too, is classified as *water*.
- Every connected component of *water* pixels is encoded as a water segment if there are more than 10 pixels in the component.

This produces a segment for most linear blue features in the map image. Figure 8.8 shows a sub-image from the image O34086E5 (rows 1820:2728, cols 879:1875); Fig. 8.9 shows the extracted linear water features.

An algorithm has been developed to find single width pixel paths through these segments as well.

Still remaining to be done is:

- segment vectorization
- determination of major branch points in linear water segments
- water feature segmentation (e.g., lake, marsh, river textures)
- blue word segmentation and interpretation.

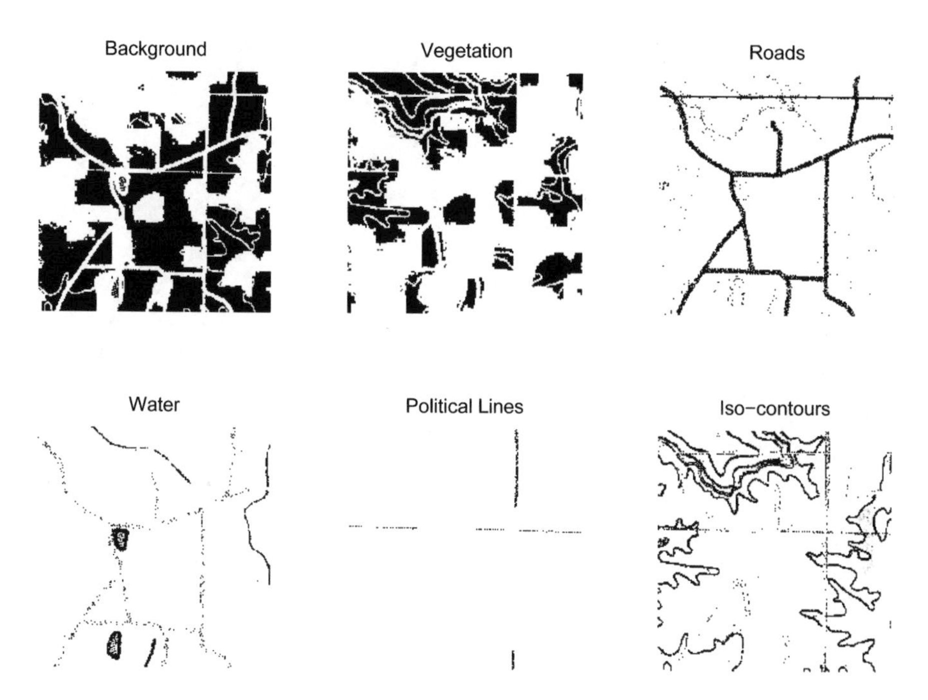

Fig. 8.6 EM segmentation results

Water Classification Results

The process described above has been applied to the set of test images, and the results are excellent.

8.5.2 Geo-political Boundary Segmentation

As described in Appendix C, there are several possible types of geo-political boundaries. Here we describe the segmentation boundaries which extend across the entire map. Geo-political lines are expected to run horizontally or vertically across the entire image (although this is not always the case). The basic algorithm at this point is:

- Find vertical and horizontal lines that have large extent.
- Adjust these lines.
- Fill in missing segments.

Figure 8.10 shows a subimage of image F34086A1 (rows 3000:3700 and cols 3000:3700), and Fig. 8.11 shows the geo-political lines segmented from it. We are currently working on extending the algorithm to work on a broader class of images.

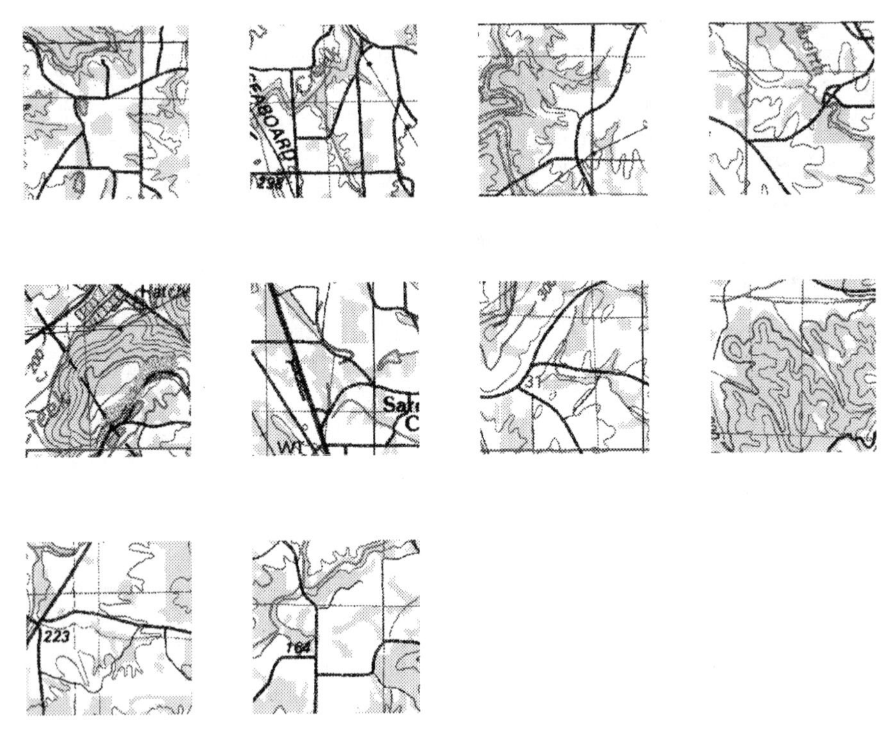

Fig. 8.7 Ten test images

8.5.3 *Iso-Contour Segmentation*

Iso-contours indicate constant level elevation in the map. Several types exist—see Appendix C. These are usually comprised of **BROWN** pixels (index equals 4 in USGS color set).

Iso-Contour Segmentation Algorithms

Currently, the iso-contour segmentation consists of simply extracting the **BROWN pixels**. Figure 8.12 shows a sub-window of image F34086A1, and Fig. 8.13 shows the extracted iso-contours.

8.5.4 *Road Marker Segmentation*

Road markers in USGS maps are:

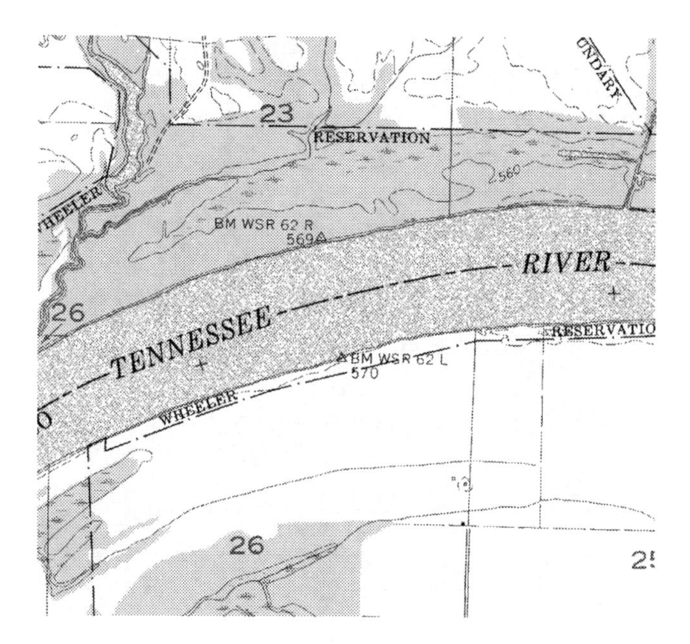

Fig. 8.8 Subimage from image O34086E5

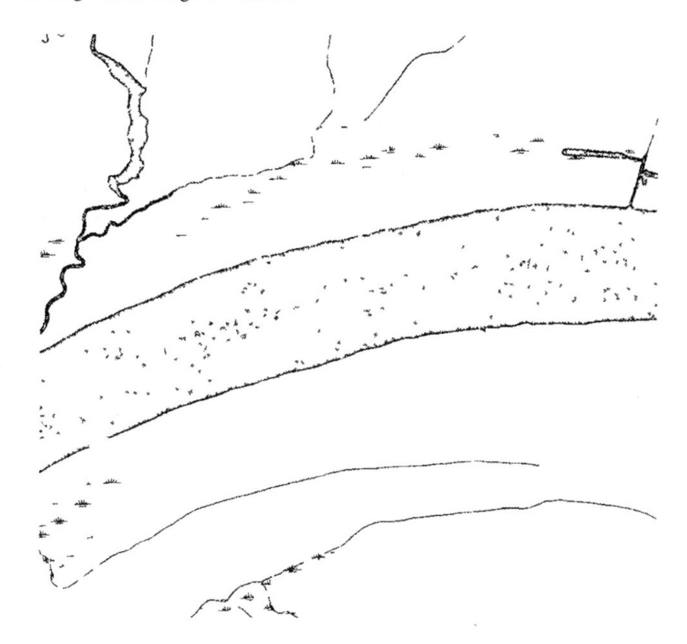

Fig. 8.9 Water features

- *State Highway Markers*: These are round or elliptical shaped (see Fig. 8.14).
- *US Highway Markers*: The form of this marker is shown in Fig. 8.15.

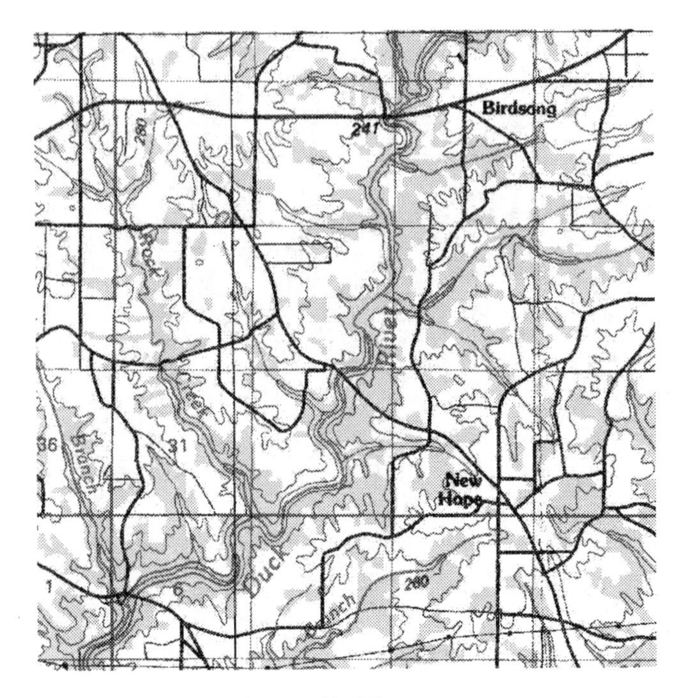

Fig. 8.10 Subimage for extraction of geo-political lines

- *Interstate Highway Markers*: The form of this marker is shown in Fig. 8.16

Figure 8.17 shows a subimage of a map with the three types of markers.

These are segmented using the Hough shape transform. Figure 8.18 shows the results on the previous image.

8.6 Texture Segmentation

Texture is used in the map to denote various classifications of land use and topography.

8.6.1 Texture Knowledge-base

The texture knowledge-base characterizes all known textures in terms of:

- Samples
- Classification Techniques
- Contextual Information.

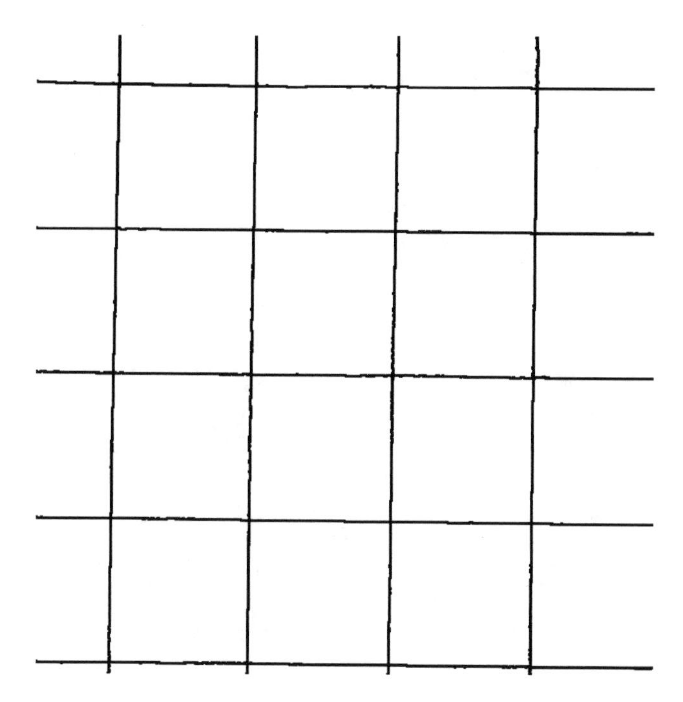

Fig. 8.11 Geo-political boundaries

8.6.2 Texture Classification Methods

Currently, textures are segmented by means of a set of sample textures. Texture features are computed on these samples, and a simple distance measure is used as a classifier.

8.6.3 Long-term Texture Classification Strategy

The long-term classification strategy in terms of adding new textures as needed, involves a semi-automated procedure:

- User designates texture regions of interest in map images in the training set of images.
- System trains automatically to obtain a classifier.
- System returns classification result.
- If accepted by user, the new texture is added to the Texture Knowledge-base.

Fig. 8.12 Subwindow of F34086A1

Fig. 8.13 Iso-contours

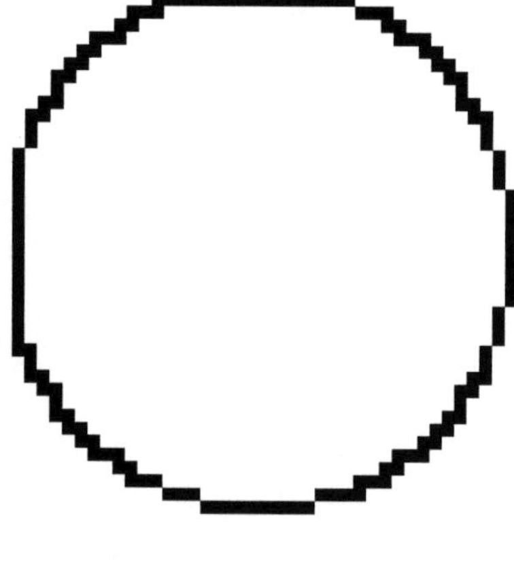

Fig. 8.14 State highway road marker

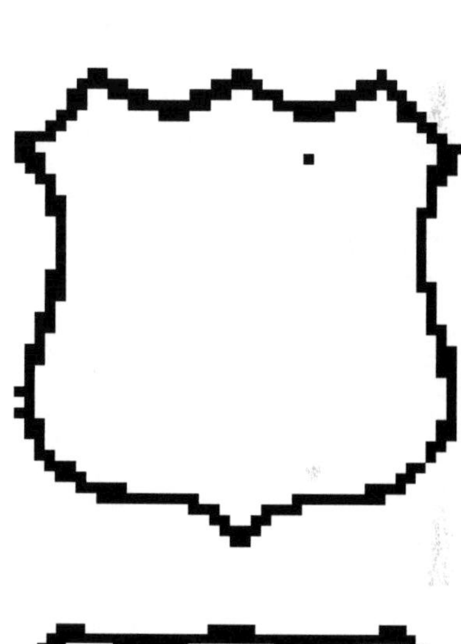

Fig. 8.15 US highway road marker

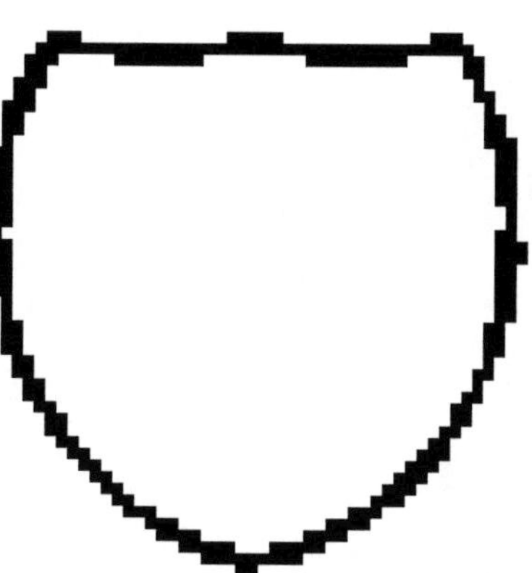

Fig. 8.16 Interstate highway road marker

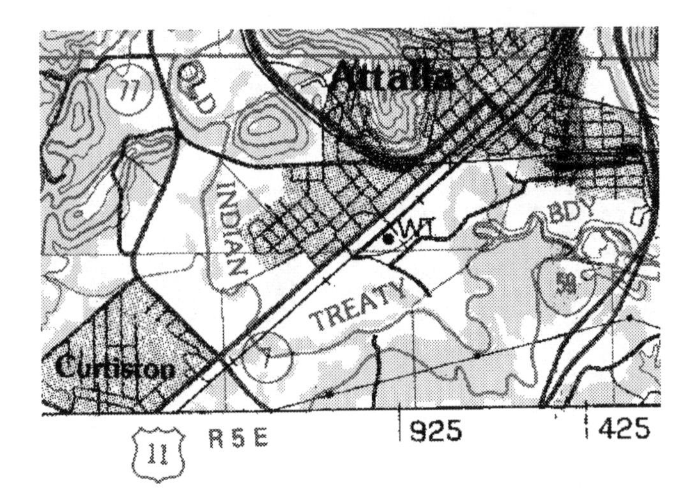

Fig. 8.17 Subimage of map with all types of road markers

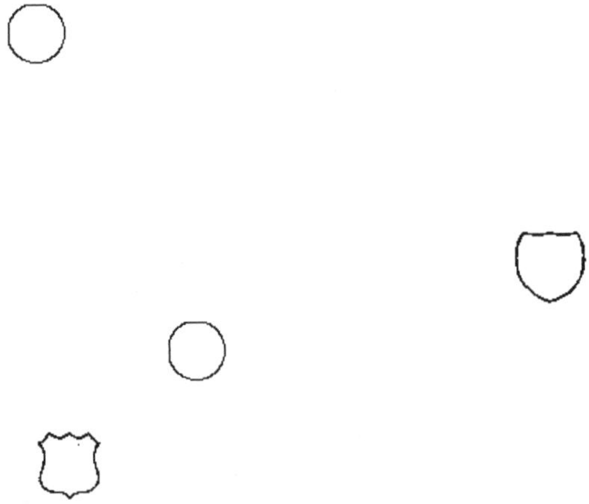

Fig. 8.18 Segmented road markers

Appendix A
Rewrite Rules for Grammar G

Rule 1
pointer_ray1 := **pointer_ray**
Rule 2
pointer_ray2 := **pointer_ray**
Rule 3
pointerarc_ray1 := **pointerarc_ray**
Rule 4
pointerarc_ray2 := **pointer_ray**
Rule 5
line_segment1 := **line_segment**
Rule 6
line_segment2 := **line_segment**
Rule 7
line_segment3 := **line_segment**
Rule 8
text1 := **text**
Rule 9
text2 := **text**
Rule 10
text_comb := *text1* + *text2*
where
near(*text1*,*text2*)
Rule 11
text_final := **text**
Rule 12
text_final := *text_comb*
Rule 13
symmetric_pointer_pair_in := *pointer_ray1* + *pointer_ray2*

T.C. Henderson, *Analysis of Engineering Drawings and Raster Map Images*,
DOI 10.1007/978-1-4419-8167-7,

where
  collinear(*pointer_ray1*,*pointer_ray2*)
  between(tail(*pointer_ray1*),head(*pointer_ray1*),head(*pointer_ray2*))
  between(tail(*pointer_ray2*),head(*pointer_ray2*),head(*pointer_ray1*))

Rule 14
symmetric_pointer_pair_out := *pointer_ray1* + *pointer_ray2*
where
  collinear(*pointer_ray1* , *pointer_ray2*)
  between(head(*pointer_ray1*),tail(*pointer_ray1*),tail(*pointer_ray2*))
  between(head(*pointer_ray2*),tail(*pointer_ray2*),tail(*pointer_ray1*))

Rule 15
symmetric_pointerarc_pair_in := *pointerarc_ray1* + *pointerarc_ray2*
where
  between(tail(*pointerarc_ray1*),head(*pointerarc_ray1*), head(*pointerarc_ray2*))
  between(tail(*pointerarc_ray2*),head(*pointerarc_ray2*),head(*pointerarc_ray1*))

Rule 16
symmetric_pointerarc_pair_out := *pointerarc_ray1* + *pointerarc_ray2*
where
  between(head(*pointerarc_ray1*),tail(*pointerarc_ray1*),tail(*pointerarc_ray2*))
  between(head(*pointerarc_ray2*),tail(*pointerarc_ray2*),tail(*pointerarc_ray1*))

Rule 17
dimension_rays_in := *symmetric_pointer_pair_in* + *text_final*
where
  near(*text_final*,head(ray1(*symmetric_pointer_pair_in*)))
  near(*text_final*,head(ray2(*symmetric_pointer_pair_in*)))

Rule 18
dimension_rays_out := *symmetric_pointer_pair_out* + *text_final*
where
  near(*text_final*,tail(ray1(*symmetric_pointer_pair_out*)))
  near(*text_final*,tail(ray2(*symmetric_pointer_pair_out*)))

Rule 19
dimension := *dimension_rays_out*

Rule 20
dimension := *dimension_rays_in*

Rule 21
dimension_angle_rays_in := *symmetric_pointerarc_pair_in* + *text_final*
where
  near(*text_final*,head(ray1(*symmetric_pointerarc_pair_in*)))
  near(*text_final*,head(ray2(*symmetric_pointerarc_pair_in*)))

Rule 22
dimension_angle_rays_out := *symmetric_pointerarc_pair_out* + *text_final*
where
  near(*text_final*,tail(ray1(*symmetric_pointerarc_pair_out*)))
  near(*text_final*,tail(ray2(*symmetric_pointerarc_pair_out*)))

Rule 23
dimension_angle := *dimension_angle_rays_out*
Rule 24
dimension_angle := *dimension_angle_rays_in*
Rule 25
dimension_set := *line_segment1* + *line_segment2* + *dimension*
where
touches(*line_segment1*,head(ray1(*dimension*)))
touches(*line_segment2*,head(ray2(*dimension*)))
perpendicular(*line_segment1*, ray1(*dimension*))
perpendicular(*line_segment2*, ray2(*dimension*))
[between(ray1(*dimension*),*line_segment1*,ray2(*dimension*))] $\wedge$
[between(*line_segment1*,ray1(*dimension*),*line_segment2*)] $\vee$
[between(*line_segment1*,ray1(*dimension*),ray2(*dimension*))] $\wedge$
[between(ray1(*dimension*),ray2(*dimension*),*line_segment2*)]
Rule 26
angle_set := *line_segment1* + *line_segment2* + *dimension_angle*
where
touches(*line_segment1*,head(ray1(*dimension_angle*)))
touches(*line_segment2*,head(ray2(*dimension_angle*)))
[between(head(ray1(*dimension_angle*)),
line_segment1,head(ray2(*dimension_angle*)))] $\wedge$
[between(head(ray1(*dimension_angle*)),
line_segment2,head(ray2(*dimension_angle*)))] $\vee$
[between(ray1(*dimension_angle*),
line_segment2,ray2(*dimension_angle*))] $\wedge$
[between(*line_segment1*, ray1(*dimension_angle*),*line_segment2*)]
Rule 27
pointer_ray_extn := **line_segment** + **pointer_ray**
where
!parallel(**line_segment**,**pointer_ray**)
horizontal(**line_segment**)
length(**line_segment**) $<$ 0.5*length(**pointer_ray**)
touches(tail(**pointer_ray**),**line_segment**)
Rule 28
pointer_line_extn := **line_segment** + **pointer_line**
where
length(**line_segment**) $>$MIN_DIST
collinear(**line_segment**,**pointer_line**)
touchesEnd(end1(**pointer_line**),**line_segment**
!touches(end2(**pointer_line**),**line_segment**)
length(**pointer_line**) + length(**line_segment**) ==
length(**pointer_line** + **line_segment**)
Rule 29
pointerarc_line_extn := **line_segment** + **pointerarc_line**

where
  touches(tail(**pointerarc_line**),**line_segment**)
  angleBetween(tail(**pointerarc_line**),**line_segment**) < 30

Rule 30

dimension_set := *line_segment1* + *line_segment2* + *pointerarc_line_extn* + *text_final*
where
  unequal(*line_segment1*,*line_segment2*)
  touches(*line_segment1*,*line_segment2*)
  !parallel(*line_segment1*,*line_segment2*)
  touches(*line_segment1*,head(ray1(*dimension*)))
  touches(*line_segment2*,tail(ray1(*dimension*)))
  near(*text_final*,ray2(*dimension*))
  below(ray2(*dimension*),*line_segment2*)
  below(ray2(*dimension*),*line_segment2*)

Rule 31

pointer_line_extn_in_circle := *pointer_line_extn* + **circle**
where
  inCircle(ray1(*pointer_line_extn*),**circle**)
  !inCircle(ray2(*pointer_line_extn*),**circle**)

Rule 32

check_sign := *line_segment1* + *line_segment2*
where
  unequal(*line_segment1*,*line_segment2*,*line_segment3*)
  length(*line_segment3*) ≥length(*line_segment1*)
  length(*line_segment3*) ≥length(*line_segment2*)
  length(*line_segment1*) ≥MIN_LENGTH
  length(*line_segment2*) ≥MIN_LENGTH
  length(*line_segment1*) <0.5*length(*line_segment2*)
  !parallel(*line_segment1*,*line_segment2*)
  !parallel(*line_segment2*,*line_segment3*)
  !parallel(*line_segment1*,*line_segment3*)
  touchesEnd(*line_segment1*,*line_segment2*)
  touchesMiddle(*line_segment1*,*line_segment3*)
  touchesMiddle(*line_segment2*,*line_segment3*)

Rule 33

check_pair := *check_sign* + *text_final*
where
  near(*text_final*,ray2(*check_sign*))
  above(*text_final*,ray2(*check_sign*))

Rule 34

dimension_description := *text_final* + *pointer_ray_extn* + **circle**
where
  near(*text_final*,tail(*poiner_ray_extn*))
  touches(**circle**,head(*pointer_ray_extn*))

!collinear(**line_segment**,head(*pointer_ray_extn*))
intersectInMiddle(**line_segment**,head(*pointer_ray_extn*))
between(*text_final*,*pointer_ray_extn*,**circle**)

Rule 35

text_in_box := **text** + **box**
where
inBox(**text**,**box**)

Rule 36

text_in_box1 = *text_in_box*

Rule 37

text_in_box2 = *text_in_box*

Rule 38

text_in_box3 = *text_in_box*

Rule 39

one_datum_ref := *text_in_box1* + *text_in_box2* + *text_in_box3*
where
unequal(*text_in_box1*,*text_in_box2*)
unequal(*text_in_box1*,*text_in_box3*)
unequal(*text_in_box2*,*text_in_box3*)
between(*text_in_box1*,*text_in_box2*,*text_in_box3*)
touchesEnd(*text_in_box1*,*text_in_box2*)
touchesEnd(*text_in_box2*,*text_in_box3*)
collinear(*text_in_box1*,*text_in_box2*)
collinear(*text_in_box2*,*text_in_box3*)

Rule 40

datum_ref := *one_datum_ref*

Rule 41

datum_below_text := *text_final* + *datum_ref*
where
near(*text_final*,*one_datum_ref*)
below(*text_final*,*one_datum_ref*)

Rule 42

dimension_description := **line_segment** + *pointer_ray_extn* + *datum_below_text*
where
length(**line_segment**) >MIN_LENGTH
!equal(**line_segment**,*pointer_ray_extn*)
!parallel(**line_segment**,ray1(*pointer_ray_extn*))
length(**line_segment**) >MIN_LENGTH
between(**line_segment**,*pointer_ray_extn*,*datum_below_text*)
touches(head(ray1(*pointer_ray_extn*)),**line_segment**)
touchesInMiddle(ray1(*pointer_ray_extn*),**line_segment**)
near(tail(ray2(*pointer_ray_extn*)),side(*datum_below_text*))

Rule 43

dimension_description := *text_final* + *pointer_line_extn_in_circle* + **line_segment**
where

horizontal(**line_segment**)
touchesEnd(**line_segment**,tail(ray2(*pointer_line_extn_in_circle*)))
near(*text_final*,**line_segment**)

Rule 44

dimension_description := *pointer_line_extn_in_circle* + *datum_below_text*
where
near(tail(ray2(*pointer_line_extn_in_circle*)),*datum_below_text*)

Rule 45

dashed_lines := *line_segment1* + *line_segment2* + *line_segment3*
where
length(*line_segment1*) ≥MIN_LENGTH
length(*line_segment3*) ≥MIN_LENGTH
unequal(*line_segment1*,*line_segment2*)
unequal(*line_segment2*,*line_segment3*)
unequal(*line_segment1*,*line_segment3*)
collinear(*line_segment1*,*line_segment2*)
collinear(*line_segment2*,*line_segment3*)
between(*line_segment1*,*line_segment2*,*line_segment3*)
near(*line_segment1*,head(*line_segment2*))
near(*line_segment3*,tail(*line_segment2*))
length(*line_segment2*) < 0.5*length(*line_segment1*)
length(*line_segment2*) < 0.5*length(*line_segment3*)
length(*line_segment1*) + length(*line_segment2*) +
length(*line_segment3*) >length(*line_segment1* + *line_segment3*)

Rule 46

dash_line1 := *dashed_lines*

Rule 47

dash_line2 := *dashed_lines*

Rule 48

circle_center_dim := *dash_line1* + *dash_line2*
where
unequal(*dash_line1*,*dash_line2*)
perpendicular(*dash_line1*,*dash_line2*)
divide(ray2(*dash_line1*),ray2(*dash_line2*))
bisect(ray2(*dash_line1*),ray2(*dash_line2*))

Rule 49

only_graphics := **graphics** + *dimension*
where
graphics != *dimension*

Rule 50

dimension_description := *dimension_set* + *only_graphics*
where
[near(head(*dimension_set*), end1(*only_graphics*))
∧ near(tail(*dimension_set*), end2(*only_graphics*))]

Appendix B
MNDAS User Manual

Thomas C. Henderson and Chimiao Xu

B.1 File and Code Organization

B.1.1 Introduction

The Matlab NonDeterministic Agent System (**MNDAS**) is an image analysis system that produces interpretations of images of engineering drawings. For example, Fig. B.1 shows a digitized CAD drawing image. The major goals of **MNDAS** are:

- to extract low level image features
  - line segments
  - arrows
  - text
  - boxes
  - circles
- to extract higher-level annotation structures:
  - dimensions
  - title block
  - materials list
  - revisions list.

More details on how these algorithms work can be found in [44, 48, 50, 113].

B.1.2 MNDAS Organization

MNDAS is a collection of Matlab functions. These must be organized as follows:

- Matlab function files are placed in a directory; e.g., let's call it: **MNDAS_main**.
- a sub-directory must be created in **MNDAS_main**; it must be called: **playpen**.

T.C. Henderson, *Analysis of Engineering Drawings and Raster Map Images*,
DOI 10.1007/978-1-4419-8167-7, © Springer Science+Business Media New York 2014

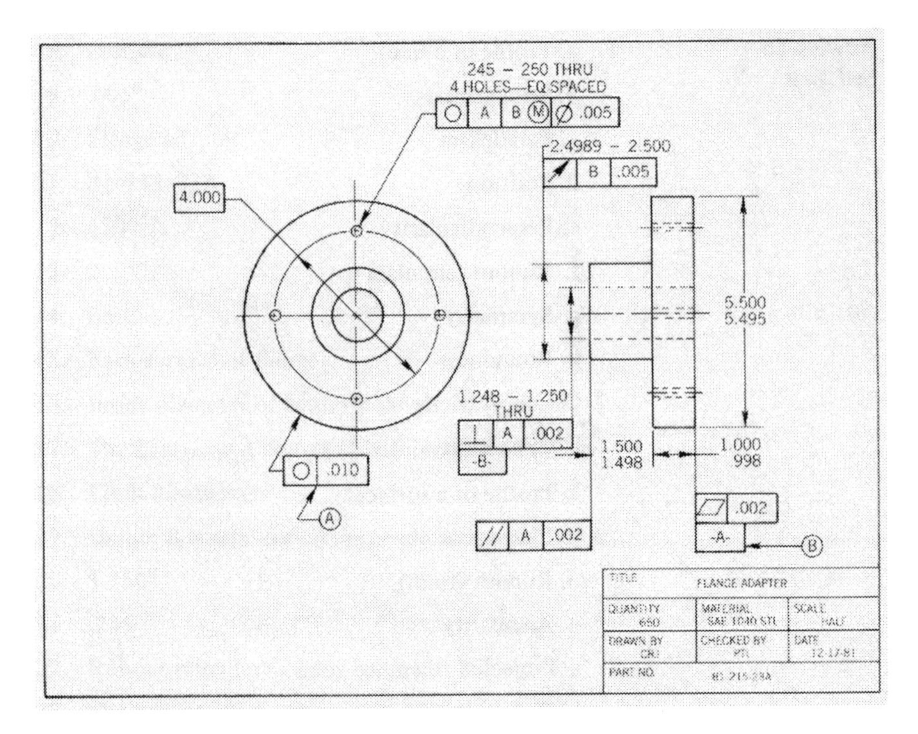

Fig. B.1 Example digital image of an engineering drawing

- In the Matlab function: *MNDAS_driver*, the global variable: *DIR_NAME* must be set to the full path name of the **playpen** directory.

Example:

We create a directory:

```
<~> iam : cd tmp
<~/tmp> iam : mkdir MNDAS_main
<~/tmp> iam : cd MNDAS_main/
<~/tmp/MNDAS_main> iam : pwd

/home/tch/tmp/MNDAS_main
```

In that directory, we create the **playpen** directory:

```
<~/tmp/MNDAS_main> iam : mkdir playpen
<~/tmp/MNDAS_main> iam : cd playpen/
<~/tmp/MNDAS_main/playpen> iam : pwd
/home/tch/tmp/MNDAS_main/playpen
<~/tmp/MNDAS_main/playpen> iam :
```

Then, we move back into the **MNDAS_main** directory:

```
<~/tmp/MNDAS_main/playpen> iam : cd ..
<~/tmp/MNDAS_main> iam :
```

Finally, we copy the Matlab **MNDAS** into the **MNDAS_main** directory:

```
<~/tmp/MNDAS_main> iam : copy <name of directory with
     files>/* .
```

B.1.3 MNDAS Execution

To run **MNDAS**, it is necessary to start Matlab in the **MNDAS_main** directory:

```
<~/tmp/MNDAS_main> iam : matlab &
```

Then in the Matlab command window, call the *MNDAS_driver* function:

```
>> MNDAS_driver
```

This starts the execution of the bf MNDAS system which will run until either:

1. a file with the name: **FC_DONE** is created in the **playpen** directory, or
2. the Matlab process is terminated.

In order to analyze an image, it is necessary to copy it into the **playpen** directory. For example:

```
copy /home/tch/tmp/drawing.tif /home/tch/tmp/
   MNDAS_main/playpen
```

Once the file is there the **MNDAS** functions will automatically work on it.
NOTE: The image file must be a TIF image with extension .tif in order for MNDAS to recognize it as an image.

B.2 Agents

B.2.1 Overview

The MNDAS framework provides a modular structure by allowing agents to be added independently of others to the greatest extent possible. However, it is still necessary for agents to understand how knowledge is communicated. The agent execution is organized as follows shown in Fig. B.2.

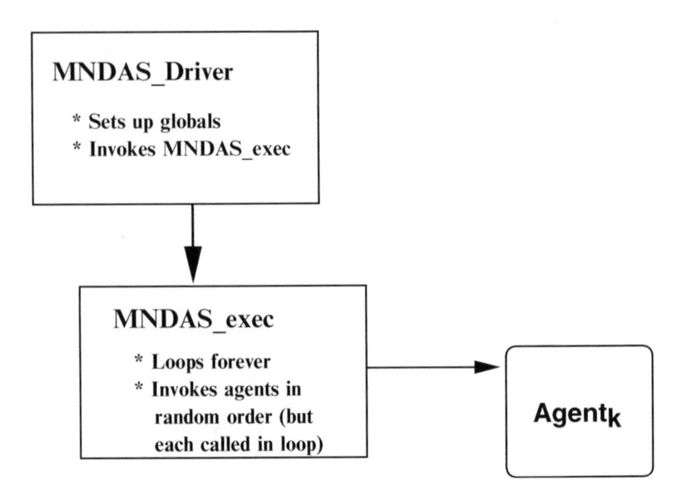

Fig. B.2 General agent architecture

Agents are divided into categories with an associated set of numbers:

- **Logistics**: agents 1–100,
- **Image Analysis**: agents 101–500,
- **Structural Analysis**: agents 501–700,
- **Miscellaneous**: agents 701–1,000,

Table B.1 gives a brief description of each agent.

Table B.1 Agent descriptions

Number	Description
1	Halt system if file **FC_done** exists
2	Provides GUI
101	Convert TIF image to base image
102	Thin binary image
105	Find branch points
104	Produce line mask image
105	Find straight line segments
120	Find circles
121	Find boxes
122	Find arrows
123	Find text

B.2.2 Logistical Agents

$Agent_1$: **Terminate**

Watches for the existence of a file named "FC_done" and if it finds it, MNDAS will halt operation.

$Agent_2$: **GUI**

Provides interactive MNDAS GUI.

B.2.3 Image Processing Agents

$Agent_{101}$: **Convert TIF to MNDAS Base Image**

Takes an arbitrary input image and, if necessary, produces a one-channel image in which the lines are darker than the background.

$Agent_{102}$: **Thin**

Thins a binary image.

$Agent_{103}$: **Branch Points**

Finds branch points in a binary image.

$Agent_{104}$: **Line Mask**

Produces line mask image.

$Agent_{105}$: **Line Segments**

Finds straight line segments in image.

$Agent_{120}$: **Circles**

Finds circles.

$Agent_{121}$: **Boxes**

Finds boxes.

$Agent_{122}$: **Arrows**

Finds arrows.

$Agent_{123}$: **Text**

Finds text.

B.2.4 Agent Communication

Upon completion of its work, an agent will create a data structure that contains information about the processing that has been performed. The agent then uses the Matlab *save* command to write the variable out to a file. The convention for file names is:

```
<filename> := a<agent_number>f<file_number>.mat
<agent_number> := current agent's number
<file_number> := integer count of files written
                 by this agent
```

E.g., $agent_{103}$ will produce files: *a103f1.mat*, *a103f2.mat*, *a103f3.mat*, etc. in that order as it works on new inputs.

Inside each file is a variable with the name: *a_variables*. Multiple items may be stored in this variable, and this is done as a vector. Each element of the vector variable has four fields:

- *data_type* (int): type of variable
- *sub_type* (int): usage of variable
- *role_type* (int): role of variable
- *value* (*data_type* specifies type): value

The *data_type* is specified by one of the following:

```
DT_IMAGE = 1;
DT_STRING = 2;
DT_FLOAT = 3;
DT_TIF_IMAGE = 4;
DT_SEGS = 5;
DT_POINTS = 6;
DT_CIRLCES = 7;
DT_BOXES = 8;
DT_PTR_RAYS = 9;
DT_TEXT = 10;
```

The *sub_type* is one of:

```
ST_THRESHOLD = 1;
ST_RAW_IMAGE = 2;
ST_INIT_IMAGE = 3;
ST_THIN_IMAGE = 4;
ST_BRANCH_IMAGE = 5;
ST_NONBRANCH_IMAGE = 6;
ST_SEGS_IMAGE = 7;
ST_SEGS_LIST = 8;
ST_POINTS_LIST = 9;
ST_STRAIGHT_SEGS_LIST = 10;
ST_CIRCLES_LIST = 11;
ST_BOXES_LIST = 12;
ST_PTR_RAY_LIST = 13;
ST_TEXT_LIST = 14;
ST_MASK_IMAGE = 15;
```

Finally, the *role_type* is on e of:

```
RT_IN = 1;
RT_OUT = 2;
RT_PARAMETER = 3;
```

An agent looks at files in the *playpen* directory, and for each unread file, the agent checks if the data type of the output variable matches the input type of the agent, it will process the data to produce a new result. Agents keep track of which files have been processed so as not to duplicate work.

Appendix C
Color Information from Legend of USGS Map

USGS maps have a well-defined structure which is exploited here to extract semantic contents. Figure C.1 shows the legend of USGS map *F34086E1.TIF*, pixels:

$$map(4001 : end - 200, 9541 : end)$$

The elements of the legend are described in terms of their location in the map image, and their constituent (non-white) pixel values. Note that the pixel values in a USGS image are coded as follows (a 5-color map uses black and the indexes 1 through 5):

Indexed Pixel Value	*Color*	*R Value*	*G Value*	*B Value*
0	Black	0	0	0
1	White	255	255	255
2	Blue	0	151	164
3	Red	203	0	23
4	Brown	131	66	37
5	Green	201	234	157
6	Purple	137	51	128
7	Yellow	255	234	0
8	Light Blue	167	226	226
9	Light Red	255	184	184
10	Light Purple	218	179	214
11	Light Gray	209	209	209
12	Light Brown	207	164	142

T.C. Henderson, *Analysis of Engineering Drawings and Raster Map Images*,
DOI 10.1007/978-1-4419-8167-7,

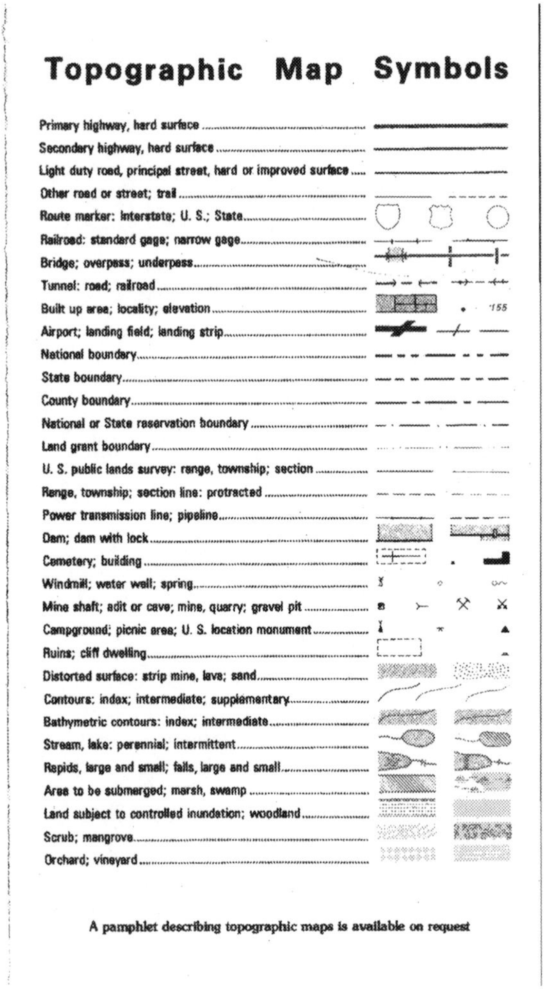

Fig. C.1 Legend of a USGS map

Topographic Map Symbols

Primary highway, hard surface
Secondary highway, hard surface
Light duty road, principal street, hard or improved surface
Other road or street; trail
Route marker: Interstate; U. S.; State
Railroad: standard gage; narrow gage
Bridge; overpass; underpass
Tunnel: road; railroad
Built up area; locality; elevation
Airport; landing field; landing strip
National boundary
State boundary
County boundary
National or State reservation boundary
Land grant boundary
U. S. public lands survey: range, township; section
Range, township; section line: protracted
Power transmission line; pipeline
Dam; dam with lock
Cemetery; building
Windmill; water well; spring
Mine shaft; adit or cave; mine, quarry; gravel pit
Campground; picnic area; U. S. location monument
Ruins; cliff dwelling
Distorted surface: strip mine, lava; sand
Contours: index; intermediate; supplementary
Bathymetric contours: index; intermediate
Stream, lake: perennial; intermittent
Rapids, large and small; falls, large and small
Area to be submerged; marsh, swamp
Land subject to controlled inundation; woodland
Scrub; mangrove
Orchard; vineyard

A pamphlet describing topographic maps is available on request

C.1 Primary Highways

Location in legend image: $legend(213:230, 718:982)$
Number of pixels of given color in feature.

Indexed Pixel Value	*Color*	Number in Subimage
0	Black	1030
1	White	2925
2	Blue	7
3	Red	456
4	Brown	352
5–12	Green	0

C.2 Secondary Highways

Location in legend image: $legend(254:270, 719:983)$
Number of pixels of given color in feature.

Indexed Pixel Value	*Color*	Number in Subimage
0	Black	533
1	White	3300
2	Blue	1
3	Red	366
4	Brown	305
5–12	Green	0

C.3 Light Duty Road

Location in legend image: $legend(300:312, 724:981)$
Number of pixels of given color in feature.

Indexed Pixel Value	*Color*	Number in Subimage
0	Black	807
1	White	2404
2	Blue	111
3	Red	0
4	Brown	12
5	Green	20
6–12		0

C.4 Other Street

Location in legend image: $legend(344:354, 722:837)$
Number of pixels of given color in feature.

Indexed Pixel Value	*Color*	Number in Subimage
0	Black	163
1	White	1060
2	Blue	45
3	Red	0
4	Brown	8
5–12		0

C.5 Other Street: Trail

Location in legend image: $legend(345:353,867:979)$
Number of pixels of given color in feature.

Indexed Pixel Value	*Color*	Number in Subimage
0	Black	137
1	White	846
2	Blue	24
3	Red	0
4	Brown	10
5–12	Green	0

C.6 Route Marker Interstate

Location in legend image: $legend(362:410,722:777)$
Number of pixels of given color in feature.

Indexed Pixel Value	*Color*	Number in Subimage
0	Black	0
1	White	2478
2	Blue	0
3	Red	159
4	Brown	107
5–12	Green	0

C.7 Route Marker US

Location in legend image: $legend(360:409,825:879)$
Number of pixels of given color in feature.

Indexed Pixel Value	*Color*	Number in Subimage
0	Black	0
1	White	2520
2	Blue	0
3	Red	147
4	Brown	83
5–12	Green	0

C.8 Route Marker State

Location in legend image: $legend(362:409, 932:982)$
Number of pixels of given color in feature.

Indexed Pixel Value	*Color*	Number in Subimage
0	Black	0
1	White	2233
2	Blue	0
3	Red	139
4	Brown	76
5–12	Green	0

C.9 Railroad: Standard

Location in legend image: $legend(423:435, 721:837)$
Number of pixels of given color in feature.

Indexed Pixel Value	*Color*	Number in Subimage
0	Black	238
1	White	1249
2	Blue	16
3	Red	0
4	Brown	18
5–12	Green	0

C.10 Railroad: Narrow

Location in legend image: $legend(421:435, 867:980)$
Number of pixels of given color in feature.

Indexed Pixel Value	*Color*	Number in Subimage
0	Black	243
1	White	1422
2	Blue	21
3	Red	0
4	Brown	24
5–12	Green	0

C.11 Bridge

Location in legend image: $legend(439:472,721:808)$
Number of pixels of given color in feature.

Indexed Pixel Value	*Color*	Number in Subimage
0	Black	308
1	White	2250
2	Blue	272
3	Red	92
4	Brown	70
5–12	Green	0

C.12 Overpass

Location in legend image: $legend(438:480,858:881)$
Number of pixels of given color in feature.

Indexed Pixel Value	*Color*	Number in Subimage
0	Black	119
1	White	747
2	Blue	2
3	Red	120
4	Brown	44
5–12	Green	0

C.13 Tunnel: Road

Location in legend image: $legend(496:514,728:850)$
Number of pixels of given color in feature.

Indexed Pixel Value	*Color*	Number in Subimage
0	Black	356
1	White	1894
2	Blue	48
3	Red	0
4	Brown	18
5	Green	21
6–12		0

C.14 Tunnel: Railroad

Location in legend image: $legend(490:511,865:982)$
Number of pixels of given color in feature.

Indexed Pixel Value	*Color*	Number in Subimage
0	Black	282
1	White	2277
2	Blue	24
3	Red	0
4	Brown	13
5–12		0

The locations of the next features are given in terms of a subimage of the legend (see Fig. C.2):

$$map(4500:5500,9541:end)$$

C.15 Builtup Area

Location in legend image: $legend(21:64,720:852)$
Number of pixels of given color in feature.

Indexed Pixel Value	*Color*	Number in Subimage
0	Black	1439
1	White	3774
2	Blue	337
3	Red	0
4	Brown	302
5–12		0

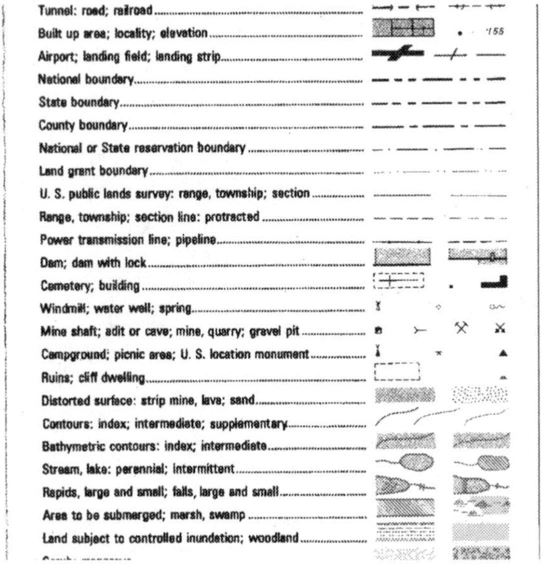

Fig. C.2 Sub-legend of a USGS map

Tunnel: road; railroad
Built up area; locality; elevation
Airport; landing field; landing strip
National boundary
State boundary
County boundary
National or State reservation boundary
Land grant boundary
U. S. public lands survey: range, township; section
Range, township; section line: protracted
Power transmission line; pipeline
Dam; dam with lock
Cemetery; building
Windmill; water well; spring
Mine shaft; adit or cave; mine, quarry; gravel pit
Campground; picnic area; U. S. location monument
Ruins; cliff dwelling
Distorted surface: strip mine, lava; sand
Contours: index; intermediate; supplementary
Bathymetric contours: index; intermediate
Stream, lake: perennial; intermittent
Rapids, large and small; falls, large and small
Area to be submerged; marsh, swamp
Land subject to controlled inundation; woodland

C.16 National Boundary

Location in legend image: $legend(127:145, 721:981)$
Number of pixels of given color in feature.

Indexed Pixel Value	*Color*	Number in Subimage
0	Black	784
1	White	4100
2	Blue	45
3	Red	0
4	Brown	30
5–12		0

C.17 State Boundary

Location in legend image: $legend(169:184, 726:981)$
Number of pixels of given color in feature.

Indexed Pixel Value	*Color*	Number in Subimage
0	Black	656
1	White	3333
2	Blue	59
3	Red	0
4	Brown	48
5–12		0

C.18 County Boundary

Location in legend image: $legend(213:226, 726:982)$
Number of pixels of given color in feature.

Indexed Pixel Value	*Color*	Number in Subimage
0	Black	650
1	White	2830
2	Blue	70
3	Red	0
4	Brown	48
5–12		0

C.19 National/State Reservation Boundary

Location in legend image: $legend(253:267, 724:981)$
Number of pixels of given color in feature.

Indexed Pixel Value	*Color*	Number in Subimage
0	Black	457
1	White	3362
2	Blue	31
3	Red	0
4	Brown	20
5–12		0

C.20 Landgrant Boundary

Location in legend image: $legend(297:309, 724:982)$
Number of pixels of given color in feature.

Indexed Pixel Value	*Color*	Number in Subimage
0	Black	0
1	White	3164
2	Blue	0
3	Red	104
4	Brown	99
5–12		0

C.21 US Public Lands Survey: Range

Location in legend image: $legend(337:352,720:840)$
Number of pixels of given color in feature.

Indexed Pixel Value	*Color*	Number in Subimage
0	Black	0
1	White	1640
2	Blue	0
3	Red	215
4	Brown	81
5–12		0

C.22 US Public Lands Survey: Section

Location in legend image: $legend(337:354,870:984)$
Number of pixels of given color in feature.

Indexed Pixel Value	*Color*	Number in Subimage
0	Black	0
1	White	1933
2	Blue	0
3	Red	74
4	Brown	63
5–12		0

C.23 Range, Township: Section Line

Location in legend image: $legend(379:396,724:844)$
Number of pixels of given color in feature.

Indexed Pixel Value	*Color*	Number in Subimage
0	Black	0
1	White	1959
2	Blue	0
3	Red	142
4	Brown	77
5–12		0

C.24 Range, Township: Protracted

Location in legend image: $legend(379:394,870:984)$
Number of pixels of given color in feature.

Indexed Pixel Value	*Color*	Number in Subimage
0	Black	0
1	White	1737
2	Blue	0
3	Red	51
4	Brown	52
5–12		0

C.25 Power Transmission Line

Location in legend image: $legend(423:435,723:841)$
Number of pixels of given color in feature.

Indexed Pixel Value	*Color*	Number in Subimage
0	Black	246
1	White	1251
2	Blue	26
3	Red	0
4	Brown	24
5–12		0

C.26 Pipeline

Location in legend image: $legend(421:433,865:981)$
Number of pixels of given color in feature.

Indexed Pixel Value	*Color*	Number in Subimage
0	Black	178
1	White	1304
2	Blue	11
3	Red	0
4	Brown	28
5–12		0

C.27 Distorted Surface: Strip Mine, Lava

Location in legend image: $legend(686:720,729:844)$
Number of pixels of given color in feature.

Indexed Pixel Value	*Color*	Number in Subimage
0	Black	4
1	White	3161
2	Blue	0
3	Red	0
4	Brown	895
5–12		0

C.28 Distorted Surface: Sand

Location in legend image: $legend(687:722,894:985)$
Number of pixels of given color in feature.

Indexed Pixel Value	*Color*	Number in Subimage
0	Black	3
1	White	2941
2	Blue	0
3	Red	0
4	Brown	368
5–12		0

C.29 Contour: Index

Location in legend image: $legend(729:764,726:819)$. Must set $(11:end,74:end)$ to 1 since there's some extra line.

Number of pixels of given color in feature.

Indexed Pixel Value	*Color*	Number in Subimage
0	Black	7
1	White	3542
2	Blue	0
3	Red	0
4	Brown	211
5–12		0

C.30 Contour: Intermediate

Location in legend image: $legend(725:759, 796:892)$. Must set $(1:14, 1:20)$ to 1 since there's some extra line.
Number of pixels of given color in feature.

Indexed Pixel Value	*Color*	Number in Subimage
0	Black	3
1	White	3253
2	Blue	0
3	Red	0
4	Brown	139
5–12		0

C.31 Contour: Supplementary

Location in legend image: $legend(723:761, 894:982)$
Number of pixels of given color in feature.

Indexed Pixel Value	*Color*	Number in Subimage
0	Black	0
1	White	3343
2	Blue	0
3	Red	0
4	Brown	128
5–12		0

C.32 Stream Lake: Perennial

Location in legend image: $legend(807:848,730:846)$
Number of pixels of given color in feature.

Indexed Pixel Value	*Color*	Number in Subimage
0	Black	28
1	White	3949
2	Blue	937
3	Red	0
4	Brown	0
5–12		0

C.33 Area to be Submerged

Location in legend image: $legend(890:927,730:844)$
Number of pixels of given color in feature.

Indexed Pixel Value	*Color*	Number in Subimage
0	Black	65
1	White	2604
2	Blue	1701
3	Red	0
4	Brown	0
5–12		0

C.34 Swamp

Location in legend image: $legend(887:923,874:985)$
Number of pixels of given color in feature.

Indexed Pixel Value	*Color*	Number in Subimage
0	Black	79
1	White	1776
2	Blue	432
3	Red	0
4	Brown	4
5	Green	1853
6–12		0

C.35 Land Subject to Controlled Inundation

Location in legend image: $legend(933:969, 732:844)$
Number of pixels of given color in feature.

Indexed Pixel Value	*Color*	Number in Subimage
0	Black	32
1	White	3027
2	Blue	1121
3	Red	0
4	Brown	1
5–12		0

C.36 Woodland

Location in legend image: $legend(930:969, 870:987)$
Number of pixels of given color in feature.

Indexed Pixel Value	*Color*	Number in Subimage
0	Black	0
1	White	1277
2	Blue	0
3	Red	0
4	Brown	0
5	Green	3443
6–12		0

The locations of the next features are given in terms of a subimage of the legend (see Fig. C.3):

$$map(5300:end-200, 9541:end)$$

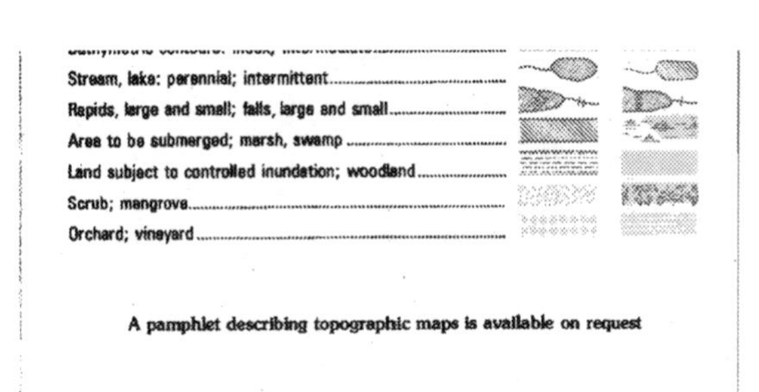

Fig. C.3 Sub-legend of a USGS map

C.37 Scrub

Location in legend image: $legend(177:212,726:845)$
Number of pixels of given color in feature.

Indexed Pixel Value	*Color*	Number in Subimage
0	Black	0
1	White	2714
2	Blue	0
3	Red	0
4	Brown	1
5	Green	1605
6–12		0

C.38 Mangrove

Location in legend image: $legend(176:212,871:985)$
Number of pixels of given color in feature.

Indexed Pixel Value	*Color*	Number in Subimage
0	Black	135
1	White	1107
2	Blue	818
3	Red	0
4	Brown	7
5	Green	2188
6–12		0

C.39 Orchard

Location in legend image: $legend(219:252,728:845)$
Number of pixels of given color in feature.

Indexed Pixel Value	*Color*	Number in Subimage
0	Black	0
1	White	2304
2	Blue	0
3	Red	0
4	Brown	0
5	Green	1706
6–12		0

C.40 Vineyard

Location in legend image: $legend(216:252, 869:983)$
Number of pixels of given color in feature.

Indexed Pixel Value	*Color*	Number in Subimage
0	Black	0
1	White	1806
2	Blue	0
3	Red	0
4	Brown	0
5	Green	2449
6–12		0

Bibliography

1. S. Ablameyko, V. Bereishik, O. Frantskevich, M. Homenko, and N. Paramonova. A System for Automatic Recognition of Engineering Drawing Entities. In *Proceedings 14th International Conference on Pattern Recognition*, pages 1157–1159, 1998.
2. S. Ablameyko and T. Pridmore. *Machine Interpretation of Line Drawing Images*. Springer Verlag, Berlin, 2000.
3. S.V. Ablameyko, V. Bereishik, O. Frantskevich, E. Mel'nik, M. Khomenko, and N. Paramonova. Interpretation of Engineering Drawings: Techniques and Experimental Results. *Pattern Recognition and Image Analysis*, 5(3):380–401, 1995.
4. S.V. Ablameyko, V. Bereishik, A. Gorelik, and S. Medvedev. 3D Object Reconstruction from Engineering Drawing Projections. *Computing and Control Engineering*, 10:277–284, 1999.
5. S.V. Ablameyko and S. Uchida. Recognition of Engineering Drawing Entities: Review of Approaches. *International Journal of Image and Graphics*, 7(4):709–733, October 2007.
6. E. Ageenko and A. Podlasov. On the Restoration of Semantic Features in Raster Topographic Images. *IADIS International Journal on Computer Science and Information Systems*, 1(1):101–114, 2006.
7. C. Ah-Soon and K. Tombre. A Step Towards Reconstruction of 3-D CAD Models from Engineering Drawings. In *Proceedings 3rd International Conference on Document Analysis and Recognition*, pages 331–334, Montral, Canada, 1995.
8. C. Ahn, K. Kim, S. Rhee, and K. Lee. A Road Extraction Method from Topographical Map Images. In *Proceedings IEEE Conference on Communications, Computers and Signal Processing*, pages 839–842, Victoria, Canada, August 1997.
9. D. Antoine, S. Colin, and K. Tombre. Analysis of Technical Documents: the REDRAW System. In *Proceedings IAPR Workshop on Structural and syntactic Pattern Recognition*, pages 192–230, New Jersey, 1990.
10. H. Arai and K. Okada. Form Processing based on background Region Analysis. In *Proceedings International Conference on Document Analysis and Recognition*, pages 164–169, Ulm, Germany, 1997.
11. N. Asher and L. Vieu. Toward a Geometry of Common Sense: A Semantics and a Complete Axiomatization of Mereotopology. In *Proceedings the International Conference on Artificial Intelligence*, Montreal, Canada, 1995.
12. F. Aurenhammer. Voronoi Diagrams: A Survey of a Fundamental Geometric Structure. *ACM Computing Surveys*, 23:345–405, 1991.
13. H.S. Baird. Background Structure in Document Images. In H. Bunke, P.S.P. Wang, and H.S. Baird, editors, *Document Image Analysis*, pages 1013–1030, Singapore, 1994. World Scientific Pub Co.

T.C. Henderson, *Analysis of Engineering Drawings and Raster Map Images*,
DOI 10.1007/978-1-4419-8167-7, © Springer Science+Business Media New York 2014

14. S. Barrat and S. Tabbone. A Bayesian Network for Combining Descriptors: Application to Symbol Recognition. *International Journal on Document Analysis and Recognition*, 13:65–75, 2009.
15. H. Blum. Biological Shape and Visual Science. *Jnl. of Theoretical Biology*, 38:205–287, 1973.
16. R. Cao and C.L. Tan. Text/Graphics Separation in Maps. In *International Workshop on Graphics Recognition*, Kingston, Canada, September 2001.
17. N. Carriero and D. Gelernter. *How to Write Parallel Programs*. MIT Press, Cambridge, 1990.
18. D.E. Catlin. *Estimation, Control, and the Discrete Kalman Filter*. Springer Verlag, New York, 1989.
19. C.-C. Chen, C.A. Knoblock, C. Shahabi, Y.-Y. Chiang, and S. Thakkar. Automatically and Accurately Conflating Orthoimagery and Street Maps. In *In Proceedings of the 12th ACM International Workshop on Geographic Information System, ACM-GIS*, pages 47–56. ACM Press, 2004.
20. A.K. Chhabra. Neural Network based Text Recognition for Engineering Drawing Conversion. In *IEEE International Conference on Neural Networks*, pages 3774–3779. IEEE, 1994.
21. Y.-Y. Chiang, C.A. Knoblock, C. Shahabi, and C.-C. Chen. Automatic and Accurate Extraction of Road Intersections from Raster Maps. *Geoinformatica*, 13(2):121–157, 2008.
22. Y.Y. Chiang. *Harvesting Geographic Features from Heterogeneous Raster Maps*. PhD thesis, Univserity of Southern California, Los Angeles, CA, December 2010.
23. S.-J. Cho and J.H. Kim. A Bayesian Network Approach for On-Line Handwriting Recognition. In B.B. Chaudhuri (Ed.), editor, *Digital Document Processing*, pages 121–141, London, UK, 2007. Springer Verlag.
24. H. Choset, K. Lynch, S. Hutchinson, G. Kantor, W. Burgard, L. Kavraki, and S. Thrun. *Principles of Robot Motion*. MIT Press, Cambridge, MA, 2005.
25. S. Collin and D. Colnet. Syntactic Analysis of Technical Drawing Dimensions. *International Journal of Pattern Recognition and Artificial Intelligence*, 8(5):1131–1148, 1994.
26. A.K. Das and N.A. Langrana. Recognition and Integration of Dimension Sets in Vectorized Engineering Drawings. *Computer Vision and Image Understanding*, 68(1):90–108, 1997.
27. S. Diana, E. Trupin, Y. Lecourtier, and J. Labiche. Document Modeling for Form Class Identification. In N.A. Murshed and F. Bortolozzi, editors, *Advances in Document Image Analysis*, pages 176–187, Berlin, 1997. Springer Verlag.
28. D. Dori. A Syntactic/Geometric Approach to Recognition of Dimensions in Engineering Drawings. *Computer Vision, Graphics and Image Processing*, 47:271–291, 1989.
29. D. Dori. Dimensioning Analysis toward Automatic Understanding of Engineering Drawings. *Communications of the ACM*, 35(10):92–103, 1992.
30. D. Dori. Orthogonal Zig-Zag: an Algorithm for Vectorizing Engineering Drawings Compared with Hough Transform. *Advances in Engineering Software*, 28:11–24, 1997.
31. D. Dori and A. Pnueli. The Grammar of Dimensions in Machine Drawings. *Computer Vision, Graphics and Image Processing*, 42:1–18, 1988.
32. D. Dori and Y. Velkovitch. Segmentation and Recognition of Dimensioning Text from Engineering Drawings. *Computer Vision and Image Understanding*, 69(2):196–201, 1998.
33. D. Dori and L. Wenyin. Vector-Based Segmentation of Text Connected to Graphics in Engineering Drawings. In *Proceedings Advances in Structural and Syntactic Pattern Recognition; LNCS 1121*, pages 322–331, Berlin, Germany, 1996. Springer Verlag.
34. D. Dori and L. Wenyin. Automated CAD Conversion with the Machine Drawings Understanding System: Concepts, Algorithms and Performance. *IEEE-T Systems, Man and Cybernetics - Part A: Systems and Humans*, 29(4):411–416, 1999.
35. L.A. Fletcher and R. Kasturi. A Robust Algorithm for Text String Seperation from Mixed Text/Graphics Images. *IEEE Transactions On Pattern Analysis And Machine Intelligence*, 10(6):910–918, November 1988.
36. Standard Upper Ontology Knowledge Interchange Format. see the formal standards document at http://suo.ieee.org/suo-kif.html.
37. D. Forsyth and J. Ponce. *Computer Vision*. Prentice Hall, Upper Saddle River, NJ, 2003.

38. H. Fuchs, Z.M. Kedem, and S.P. Uselton. Optimal Surface Reconstructions from Planar Contours. *Communications of the ACM*, 20(10):693–702, 1977.
39. B. Grocholsky, A. Makarenko, T. Kaupp, and H. Durrant-Whyte. Scalable Control of Decentralised Sensor Platforms. In *Proceedings of Information Processing in Sensor Networks*, pages 96–112, Palo Alto, CA, 2003.
40. D.S. Guru, B.H. Shekar, and P. Nagabhushan. A Simple and Robust Line Detection Algorithm based on Eigenvaluee Analysis. *Pattern Recognition Letter*, 25(1):1–13, 2004.
41. T. Ha and H. Bunke. Model-based Analysis and Understanding of Check Forms. In H. Bunke, P.S.P. Wang, and H.S. Baird, editors, *Document Image Analysis*, pages 57–84, Singapore, 1994. World Scientific Pub Co.
42. A. Habed and B. Boufama. Dimension Sets in Technical Drawings. In *Proceedings of Vision Interface*, pages 217–223, Trios-Rivieres, CA, 1999.
43. T.C. Henderson. *Discrete Relaxation*. Oxford University Press, Oxford, 1990.
44. T.C. Henderson. Explicit and Persistent Knowledge in Engineering Drawing Analysis. Research Report UUCS-03-018, School of Computing, University of Utah, Salt Lake City, Utah, October 2003.
45. T.C. Henderson and T. Linton. Raster Map Image Analysis. In *Poster Session of International Conference on Document Analysis and Recognition*, Catalonia, Spain, July 2009.
46. T.C. Henderson, T. Linton, S. Potupchik, and A. Ostanin. Automatic Segmentation of Semantic Classes in Raster Map Images. In *IARP International Workshop on Graphics Recognition*, La Rochelle, France, July 2009.
47. T.C. Henderson and L. Swaminathan. Agent-Based Engineering Drawing Analysis. Research Report UUCS-02-008, School of Computing, University of Utah, Salt Lake City, Utah, February 2002.
48. T.C. Henderson and L. Swaminathan. Form Analysis with the Nondeterministic Agent System (NDAS). In *Proceedings of 2003 Symposium on Document Image Understanding Technology*, pages 253–258, April 2003.
49. T.C. Henderson and L. Swaminathan. NDAS: The Nondeterministic Agent System for Engineering Drawing Analysis. In *Proceedings Intl Conference on Integration of Knowledge Intensive Multi-agent Systems*, pages 512–516, Boston, MA, October 2003.
50. T.C. Henderson and L. Swaminathan. Symbolic Pruning in a Structural Approach to Engineering Drawing Analysis. In *Proceedings Intl Conference on Document Analysis and Recognition*, pages 180–184, Edinburgh, Scotland, August 2003.
51. T.C. Henderson and C. Xu. Robot Navigation Techniques for Engineering Drawing Analysis. In *Proceedings of 2005 Symposium on Document Image Understanding Technology*, pages 201–210, College Park, MD, November 2005.
52. Thomas C. Henderson and Ashok Samal. Shape Grammar Compilers. *Pattern Recognition*, 19(4):279–288, 1985.
53. X. Hilaire and K. Tombre. Robust and Accurate Vectorization of Line Drawings. *IEEE-T on Pattern Analysis and Machine Intelligence*, 28(6):890–904, June 2006.
54. S. Hinz and A. Baumgartner. Automatic Extraction of Urban Road Networks from Multi-view Aerial Imagery. *ISPRS Journal of Photogrammetry and Remote Sensing*, 58(1–2):83–98, June 2003.
55. T.V. Hoang and S. Tabbone. Text Extraction from Graphical Document Images using Sparse Representation. In *Proceedings Intl Workshop on Document Analysis Systems*, pages 143–150, Boston, MA, June 2010.
56. K. Hormann, S. Spinello, and P. Schröder. C1-Continuous Terrain Reconstruction from Sparse Contours. In *Proceedings of Vision Modeling and Visualization*, pages 289–297, München, Germany, November 2005.
57. M.K. Hu. Visual Pattern Recognition by Moment Invariants. *IRE Transactions on Information Theory*, IT-8:179–187, 1962.
58. W. Itonaga, I. Matsuda, N. Yoneyama, and S. Ito. Automatic Extraction of Road Networks from Map Images. *Electronics and Communications in Japan*, 86(4):62–72, 2003.

59. S.H. Joseph and T.P. Pridmore. Knowledge-Directed Interpretation of Mechanical Engineering Drawings. *IEEE-T Pattern Analysis and Machine Intelligence*, 14(9):928–940, 1992.
60. S.H. Joseph, T.P. Pridmore, and M.E. Dunn. Toward the Automatic Interpretation of Mechanical Engineering Line Drawings. In A. Bartlett, editor, *Computer Vision and Image Processing*, New York, NY, 1989. Kogan Page.
61. Y.-B. Kang, S.-Y. Ok, and H.-G. Cho. Character Grouping Technique using 3-D Neighborhood Graphs in Raster Map. In *Proceedings of the 14th International Conference on Pattern Recognition*, Brisbane, Australia, August 1998.
62. T. Kanungo, R. Haralick, and D. Dori. Understanding Engineering Drawings: A Survey. In *Proceedings of First IARP Workshop on Graphics Recognition*, pages 217–228, University Park, PA, 1995.
63. A. Khotanzad and E. Zink. Color Map Segmentation using Eigenvector Line-Fitting. In *Proceedings of the IEEE Southwest Symposium on Image Analysis and Interpretation*, San Antonio, TX, April 1996.
64. A. Khotanzad and E. Zink. Contour Line and Geographic Feature Extraction from USGS Color Topographical Paper Maps. *IEEE Transactions on Pattern Analysis and Machine Intelligence*, 25(1):18–31, 2003.
65. C.P. Lai and R. Kasturi. Detection of Dimension Sets in Engineering Drawings. *IEEE-T on Pattern Analysis And Machine Intelligence*, 16(8):848–855, August 1994.
66. L. Lam, S.-W. Lee, and C.Y. Suen. Thinning Methodologies – A Comprehensive Survey. *IEEE-T on Pattern Analysis And Machine Intelligence*, 14(9):869–885, September 1992.
67. J.J. Lee, J. Kim, and J.H. Kim. Data-Driven Design of HMM Topology for Online Handwriting Recognition. In H. Bunke and T. Caelli, editors, *Hidden Markov Models: Applications in Computer Vision*, pages 107–121, Singapore, 2001. World Scientific Pub Co.
68. S. Levachkine, A. Velázquez, V. Alexandrov, and M. Kharinov. Semantic Analysis and Recognition of Raster-Scanned Color Cartographic Images. In *Selected Papers from the Fourth International Workshop on Graphics Recognition Algorithms and Applications*, GREC '01, pages 178–189, London, UK, 2002. Springer Verlag.
69. L. Li, G. Nagy, A. Samal, S. Seth, and Y. Xu. Cooperative Text and Line-art Extraction from a Topographic Map. In *Proceedings of the Fifth International Conference on Document Analysis and Recognition*, Bangalore, India, September 1999.
70. L. Li, G. Nagy, A. Samal, S. Seth, and Y. Xu. Integrated Text and Line-Art Extraction from a Topographic Map. *International Journal on Document Analysis and Recognition*, 2:175–185, 2000.
71. J. Liang, J. Ha, R.M. Haralick, and I.T. Phillips. Document Layout Structure Extraction using Bounding Boxes of Different Entities. In *Proceedings 3rd IEEE Workshop on Applications of Computer Vision (WACV '96)*, pages 278–283, December 1996.
72. T. Linton. Semantic Feature Analysis in Raster Maps. Master's thesis, University of Utah, Salt Lake City, Utah, August 2009.
73. W. Liu and D. Dori. Sparse Pixel Tracking: A Fast Vectorization Algorithm Applied to Engineering Drawings. In *Proceedings of the 13th International Conference on Pattern Recognition*, pages 808–812, Vienna, Austria, August 1996. IEEE.
74. W. Liu and D. Dori. Genericity in Graphics Recognition Algorithms. In K. Tombre and A. Chhabra (Eds.), editors, *Graphics Recognition (GREC) 1998*, LNCS 1389, pages 9–21, Berlin, Germany, 1998. Springer Verlag.
75. W. Liu, W. Zhang, and L. Yan. An Interactive Example-Driven Approach to Graphics Recognition in Engineering Drawings. *International Journal on Document Analysis and Recognition*, 9:13–29, 2007.
76. J. Lladós. Advances in Graphics Recognition. In B.B. Chaudhuri (Ed.), editor, *Digital Document Processing*, pages 281–303, London, UK, 2007. Springer Verlag.
77. T. Lu, C.-L. Tai, H. Yang, and S. Cai. A Novel Knowledge-Based System for Interpreting Complex Engineering Drawings: Theory, Representation, and Implementation. *IEEE Transactions on Pattern Analysis and Machine Intelligence*, 31(8):1444–1457, 2009.

78. Z. Lu. Detection of Text Regions from Digital Engineering Drawings. *IEEE Transactions on Pattern Analysis and Machine Intelligence*, 20(4):431–439, 1998.
79. S. Mandal, A.K. Das, and P. Bhowmick. A Fast Technique for Vectorization of Engineering Drawings using Morphology and Digital Straightness. In *ACM Proceedings of the International Conference on Computer Vision, Graphics and Image Processing*, Chennai, India, December 2010.
80. S. Mao and T. Kanungo. Empirical Performance Evaluation Methodology and Its Application to Page Segmentation Algorithms. *IEEE Transactions on Pattern Analysis and Machine Intelligence*, 23(3):242–256, 2001.
81. D. Marr. *Vision*. W.H. Freeman and Company, New York, NY, 1982.
82. S. Marsland. *Machine Learning, An Algorithmic Approach*. Chapman & Hall, CRC Press, Boca Raton, FL, 2009.
83. M. Mattavelli, V. Noel, and E. Amaldi. A New Approach for Fast Line Detection Based on Combinatorial Optimization. In *Proceedings of the 10th International Conference on Image Analysis and Processing*, pages 168–173, Venice, Italy, 1999. IEEE.
84. G. Medioni and S. Kang. *Emerging Topics in Computer Vision*. Prentice Hall, Boston, MA, 2004.
85. G. Medioni, M. Lee, and C. Tang. *Computational Framework for Segmentation and Grouping*. Elsevier Science Inc., New York, NY, 2000.
86. T. Miyoshi, W. Li, K. Kaneda, and E. Nakamae. Automatic extraction of buildings utilizing geometric features of a scanned topographic map. *Proceedings of the 17th International Conference on Pattern Recognition*, pages 626–629, Washington, DC, 2004. IEEE.
87. R. Mohr and T.C. Henderson. Arc and Path Consistency Revisited. *Artificial Intelligence*, 28(2):225–233, March 1986.
88. S. Mori, H. Nishida, and H. Yamada. *Optical Character Recognition*. John Wiley and Sons, Hoboken, NJ, 1999.
89. L. Najman, O. Gibot, and S. Berche. Indexing Technical Drawings using Title Block Structure Recognition. In *Proceedings 6th International Conference on Document Analysis and Recognition*, Seattle, WA, 2001.
90. I. Niles and A. Pease. Towards a Standard Upper Ontology. In *Proceedings of the 2nd International Conference on Formal Ontology and Information Systems*, 2001.
91. H. Nishida and S. Mori. A Model-based Split-and-Merge Method for Character String Recognition. *International Journal of Pattern Recognition and Artificial Intelligence*, 8(5):1205–1222, 1994.
92. L. O'Gorman and R. Kasturi. *Document Image Analysis*. IEEE Press, Piscataway, NJ, 1997.
93. M. Ondrejcek, J. Kastner, R. Kooper, and P. Bajcsy. Information Extraction from Scanned Engineering Drawings. Image Spatial Analysis Group NCSA-ISDA09-001, National Center for Supercomputing Applications, December 2009.
94. A. Pezeshk and R.L. Tutwiler. Contour Line Recognition and Extraction from Scanned Color Maps using Dual Quantization of the Intensity Image. In *Proceedings of the IEEE Southwest Symposium on Image Analysis and Interpretation*, Sante Fe, NM, March 2008.
95. A. Podlasov, E. Ageenko, and P. Fränti. Morphological Recontruction of Semantic Layers in Map Images. *Journal of Electronic Imaging*, 15(1), March 2006.
96. J. Pouderoux, J.C. Gonzato, A. Pereira, and P. Guitton. Toponym Recognition in Scanned Color Topographic Maps. In *Proceedings International Conference on Document Analysis and Recognition*, pages 531–535, Parana, Brazil, September 2007.
97. C. Poullis, S. You, and U. Neumann. A Vision-Based Systenm for Automatic Detection and Extraction of Road Networks. In *Proceedings of the 2008 IEEE Workshop on Applications of Computer Vision*, pages 1–8, Washington, DC, 2008. IEEE.
98. B.S. Prabhu and S.S. Pande. Intelligent Interpretation of CADD Drawings. *Computers and Graphics*, 23:25–44, 1999.
99. I. Pratt and O. Lemon. Ontologies for Plane, Polygonal Mereotopology. Technical Report UMCS-97-1-1, University of Manchester, 1997.

100. P.P. Roy, E. Vazquez, J. Lladós, R. Baldrich, and U. Pal. A System to Segment Text and Symbols from Color Maps. In W. Liu, J. Lladós, and J.-M. Ogier (Eds), editors, *Graphics Recognition (GREC) 2007*, LNCS 5046, pages 245–256, Berlin, Germany, 2008. Springer Verlag.
101. S. Russell and P. Norvig. *Artificial Intelligence*. Prentice-Hall, Upper Saddle River, NJ, 2003.
102. Spinello Salvatore and Pascal Guitton. Contour Line Recognition From Scanned Topographic Maps. In *WSCG (Winter School of Computer Graphics)*, 2004.
103. R. Samet, I.N. Askerzade Askerbeyli, and C. Varol. An Implementation of Automatic Contour Line Extraction from Scanned Digital Topographic Maps. *Appl. Comput. Math.*, 9:116–127, 2010.
104. L.M. San, S.M. Yatim, N.A.M. Sheriff, and N. Isrozaidi. Extracting Contour Lines from Scanned Topographic Maps. In *Proceedings of the 2008 IEEE Conference on Computer Graphics, Imaging and Visualization*, pages 187–192, Washington, DC, July 2004. IEEE.
105. X. Shao, L. Ye, M. Cai, and Y. Wang. Simultaneous Image De-Noising and Curve Extraction by Tensor Voting. In *Proceedings of the 2008 Intl Symposium on Computer Science and Computational Technology*, pages 536–538, Hong Kong, China, December 2008.
106. W. Shen, X. Bai, R. Hu, H. Wang, and L.J. Latecki. Skeleton Growing and Pruning with Bending Potential Ratio. *Pattern Recognition*, 44(2):196–209, February 2011.
107. B.-S. Shin and H.-S. Jung. Contour-based Terrain Model Reconstruction using Distance Information. In *Proceedings Computational Science and Its Applications (ICCSA)*, pages 561–585, Singapore, May 2005.
108. J. Song, M. Cai, M.R. Lyu, and S. Cai. A New Approach for Line Recognition in Large-size Images using Hough Transform. In *Proceedings of the Internatoinal Conference on Pattern Recognition*, pages 33–36, 2002.
109. J. Song, M.R. Lyu, and S. Cai. Effective Multiresolution Arc Segmentation: Algorithms and Performance Evaluation. *IEEE Transactions on Pattern Analysis and Machine Intelligence*, 26(11), 2004.
110. J. Song, F. Su, C.L. Tai, and S. Cai. An Object-Oriented Progressive-Simplification-Based Vectorization System for Engineering Drawings: Model, Algorithm, and Performance. *IEEE Transactions on Pattern Analysis and Machine Intelligence*, 24(8), 2002.
111. F. Su, J. Song, C.-L. Tai, and S. Cai. Dimension Recognition and Geometry Reconstruction in Vectorization of Engineering Drawings. In *Proceedings of Intl Conference on Computer Vision and Pattern Recognition*, pages 710–716, 2001.
112. V.S. Subrahmanian. *Heterogeneous Agent Systems*. MIT Press, Cambridge, MA, 2000.
113. L. Swaminathan. Agent-Based Engineering Drawing Analysis. Master's thesis, University of Utah, Salt Lake City, Utah, May 2003.
114. T. Syeda-Mahmood. Extracting Indexing Keywords from Image Structures in Engineering Drawings. In *Proceedings 5th International Conference on Document Analysis and Recognition*, pages 471–474, Bangalore India, 1999.
115. S. Tabbone and L. Wendling. Technical Symbols Recognition using the Two-Dimensional Radon Transform. In *Proceedings of Intl Conference on Pattern Recognition*, volume 3, Washington, DC, 2002. IEEE.
116. Y. Tang and J. Lin. Information Acquisition and Storage of Forms in Document Processing. In *Proceedings International Conference on Document Analysis and Recognition*, pages 170–174, Ulm, Germany, 1997.
117. A. Tarski. Foundations of the Geometry of Solids. In *Logic, Semantics, Mathematics*, pages 24–30, Oxford, UK, 1956. Oxford University Press.
118. K. Tombre. Analysis of Engineering Drawings: State of the Art and Challenges. In *In Graphics Recognition: Algorithms and Systems*, volume 1389 of *Lecture Notes on Computer Science*, pages 257–264, Berlin, 1998. Springer Verlag.
119. K. Tombre. Graphics Documents: Achievements and Open Problems. In *Proceedings of 10th Portuguese Conference on Pattern Recognition*, Portugal, 1998.

120. K. Tombre and D. Antoine. Analysis of Technical Documents using A Priori Knowledge. In *Proceedings IARP Workshop on Syntactic and Structural Pattern Recognition*, pages 178–189, Pont-Mousson, France, 1988.
121. K. Tombre and D. Dori. Interpretation of Engineering Drawings. In H. Bunke and P.S.P. Wang, editors, *Handbook of Character Recognition and Document Image Analysis*, pages 457–484, Singapore, 1997. World Scientific Pub Co.
122. K. Tombre, S. Tabbone, L. Pélissier, B. Lamiroy, and P. Dosch. Text/Graphics Separation Revisited. In *DAS 2002 Proceedings of the 5th International Workshop on Document Analysis Systems*, pages 200–211, London, UK, 2002. Springer Verlag.
123. P. Vaxiere and K. Tombre. Celesstin: CAD Conversion of Mechanical Drawings. *IEEE Computer*, 25(7):46–54, 1992.
124. A. Veláquez and S. Levachkine. Text/Graphics Separation and Recognition in Raster-scanned Color Cartographic Maps. In *Proceedings of the 2003 Workshop on Graphics Recognition*, Barcelona, Spain, July 2003.
125. D. Wang and S.N. Srihari. Analysis of Form Images. *International Journal of Pattern Recognition and Artificial Intelligence*, 8(5):1031–1051, 1994.
126. Q. Wang, J. Shi, and D.D. Feng. A Uniform Framework of Representation and Structure Reconstruction for Generic Form Image. In *Proceedings 7th International Conference on Signal Processing (ICSP)*, pages 1052–1055, August 2004.
127. Y. Wang, I.T. Phillips, and R.M. Haralick. Using Area Voronoi Tessellation to Segment Characters Connected to Graphics. In *Proceedings of the 2001 Workshop on Graphics Recognition*, Kingston, Canada, September 2001.
128. Y. Wang, L. Tang, and Z. Tang. A New Method to Recognize Dimension Sets and its Application in Architectural Drawings. In *Proceedings of the 5th International Conference on Computer Aided Design and Computer Graphics*, Shenzhen, China, December 1997.
129. G. Weiss. *Multi-Agent Systems*. MIT Press, Cambridge, MA, 1999.
130. P. Winston. *Artificial Intelligence*. Addison-Wesley, Reading, MA, 1984.
131. C. Xu. The Analysis of Engineering Drawings using Robot Mapping Techniques. Master's thesis, University of Utah, Salt Lake City, Utah, August 2006.
132. Y.-Y.Chiang and C.A. Knoblock. Classification of Line and Character Pixels on Raster Maps using Discrete Cosine Transformation Coefficients and Support Vector Machines. In *Proceedings Intl Conference on Pattern Recognition*, Washington, DC, USA, August 2006. IEEE Computer Society.
133. Y.-Y.Chiang and C.A. Knoblock. Automatic Extraction of Road Intersection Position, Connectivity and Orientations from Raster Maps. In *Proceedings Intl Conference on Advances in Geographic Information Systems*, November 2008.
134. Y.-Y.Chiang and C.A. Knoblock. An Approach for Recognizing Text Labels in Raster Maps. In *Proceedings Intl Conference Pattern Recognition*, August 2010.
135. Y.-Y.Chiang, C.A. Knoblock, and C.-C. Chen. Automatic Extraction of Road Intersections from Raster Maps. In *Proceedings Intl Conference on GIS*, Bremen, Germany, November 2005.
136. H. Yamada. Paper-based Map Processing. In H. Bunke and P.S.P. Wang, editors, *Handbook of Character Recognition and Document Image Analysis*, pages 503–528, Singapore, 1997. World Scientific Pub Co.
137. H. Yamada, K. Yamamoto, T. Saito, and K. Hosokawa. Recognition of Elevation Value in Topographic Maps by Multi-Angled Parallelism. *International Journal of Pattern Recognition and Artificial Intelligence*, 8(5):1149–1170, 1994.
138. Y. Yu, A. Samal, and S.C. Seth. A System for Recognizing a Large Class of Engineering Drawings. *IEEE Transactions on Pattern Analysis and Machine Intelligence*, 19(8): 868–890, 1997.
139. Y.H. Yu, A. Samal, and S.C. Seth. Isolating Symbols from Connection Lines in a Class of Engineering Drawings. *Pattern Recognition*, 27(3):391–404, March 1994.
140. S. Zheng, J. Liu, W. Shi, and G. Shu. Road Central Contour Extraction from High Resolution Satellite Image using Tensor Voting Framework. In *Proceedings of the 5th Intl Conference on Machine Learning and Cybernetics*, pages 3248–3253, Guangzhou, China, August 2006.

Index

T.C. Henderson, *Analysis of Engineering Drawings and Raster Map Images*,
DOI 10.1007/978-1-4419-8167-7,

Printed in the United States
By Bookmasters